3044910005052
Sigl, Mark A.
Mechanical drawing problems
workbook for Autocad : Release 12

# Mechanical Drawing Problems Workbook For AutoCAD

## Release 12

# Mechanical Drawing Problems Workbook For AutoCAD Release 12

by

Mark A. Sigl

WEST GEORGIA TECHNICAL COLLEGE
LIBRARY

Delmar Publishers Inc.

## NOTICE TO THE READER

Publisher does not warrant or guarantee any of the products described herein or perform any independent analysis in connection with any of the product information contained herein. Publisher does not assume, and expressly disclaims, any obligation to obtain and include information other than that provided to it by the manufacturer.

The reader is expressly warned to consider and adopt all safety precautions that might be indicated by the activities described herein and to avoid all potential hazards. By following the instructions contained herein, the reader willingly assumes all risks in connection with such instructions.

The publisher makes no representations or warranties of any kind, including but not limited to, the warranties of fitness for particular purpose or mechantability, nor are any such representations implied with respect to the material set forth herein, and the publisher takes no responsibility with respect to such material. The publisher shall not be liable for any special, consequential or exemplary damages resulting, in whole or in part, from the readers' use of, or reliance upon, this material.

Trademarks

AutoCAD® is registered in the U.S. Patent and Trademark Office by Autodesk, Inc.

For information, address Delmar Publishers Inc.
3 Columbia Circle, Box 15-015
Albany, NY 12212-5015

Copyright © 1993 by Delmar Publishers Inc.
All rights reserved. No part of this work may be reproduced or used in any form, or by any means — graphic, electronic, or mechanical, including photocopying, recording, taping, or information storage and retrieval systems — without written permission of the publisher.

Printed in the United States of America
Published simultaneously in Canada
by Nelson Canada,
a Division of the Thomson Corporation

1  2  3  4  5  6  7  8  9  10  XXX  99  98  97  96  95  94  93

ISBN: 0-8273-4829-0

# CONTENTS

| PROBLEM | | PAGE |
|---|---|---|
| **CHAPTER 1 BASIC DRAWING COMMANDS** | | |
| | INTRODUCTION | 1 |
| 1 | BORDER PROTYPE DRAWING | 2, 3 |
| 2 | LINES AND LAYERS | 4 |
| 3 | SNAP AND GRID | 4 |
| 4 | ABSOLUTE COORDINATES | 5 |
| 5 | RELATIVE COORDINATES | 6 |
| 6 | POLAR COORDINATES | 7 |
| 7 | CIRCLES AND OSNAP | 8 |
| 8 | ARC AND DONUT | 9 |
| 9 | DTEXT & TEXT | 10 |
| 10 | POLYGON & OFFSET | 11, 12 |
| | | 13 |
| **CHAPTER 2 BASIC EDITING COMMANDS** | | |
| | INTRODUCTION | 14 |
| 11 | RECTANGULAR ARRAY | 15 |
| 12 | POLAR ARRAY | 16 |
| 13 | TRIM, ARRAY, POLYGON | 17 |
| 14 | COPY | 18 |
| 15 | CHANGE | 19 |
| 16 | CHAMFER | 20 |
| 17 | FILLET | 21 |
| 18 | MIRROR | 22 |
| 19 | PLINE, PEDIT & STRETCH | 23, 24 |
| 20 | SCALE AND ROTATE | 25 |
| **CHAPTER 3 BLOCKS, ATTRIBUTES, AND SCALE FACTORS** | | |
| | INTRODUCTION | 26 |
| 21 | ELECTRICAL BLOCK LIBRARY | 27 |
| 22 | ELECTRICAL SCHEMATIC | 28 |
| 23 | TITLE BOX ATTRIBUTE AND BORDER UPDATE | 29, 30 |
| 24 | NESTED BLOCKS | 31 |
| 25 | PARTS LIST WITH ATTRIBUTES | 32 |
| 26 | BORDER "C" SIZE | 33, 34 |
| 27 | SPLIT GLAN | 35 |
| 28 | PUMP HOUSING HALF-SIZE DRAWING | 36 |
| 29 | HING PLATE | 37 |
| 30 | STOP PLATE | 38 |
| **CHAPTER 4 DIMENSIONING** | | |
| | INTRODUCTION | 39 |
| 31 | DIM VARIABLES | 40 |
| 32 | HORIZONTAL AND VERTICAL | 41 |
| 33 | LEADER, DIAMETER, RADIUS | 42 |
| 34 | BASELINE | 43 |
| 35 | CONTINUOUS | 44 |
| 36 | ALIGN AND ROTATE | 45 |
| 37 | LIMITS TOLERANCING | 46 |
| 38 | TOLERANCE | 47 |
| 39 | RAM HEAD | 48 |
| 40 | FRACTIONAL | 49 |
| **CHAPTER 5 ORTHOGRAPHIC** | | |
| | INTRODUCTION | 50 |
| 41 | BASE BRACKET | 51 |
| 42 | BEARING BRACKET | 52 |
| 43 | SLIDE BRACKET | 53 |
| 44 | SUPPORT BRACE | 54 |
| 45 | T-GUIDE | 55 |
| 46 | UNIVERSAL JOINT | 56 |

| PROBLEM | | PAGE |
|---|---|---|

**CHAPTER 5 continued**

| 47 | BEARING SADDLE | 57 |
| 48 | SHAFT SET | 58 |
| 49 | COLUMN BASE | 59 |
| 50 | WING POST | 60 |

**CHAPTER 6 ISOMETRIC AND OBLIQUE**

| | INTRODUCTION | 61 |
| 51 | T-GUIDE | 62 |
| 52 | STOP BRACKET | 63 |
| 53 | FIXTURE END BLOCK | 64 |
| 54 | HOLD-DOWN CLAMP | 65 |
| 55 | SLIDE BLOCK | 66 |
| 56 | LUG | 67 |
| 57 | SHAFT SET | 68 |
| 58 | LIFT BLOCK | 69 |
| 59 | UNIVERSAL JOINT | 70 |
| 60 | BEARING BLOCK | 71 |

**CHAPTER 7 SECTIONS AND HATCHING PROBLEMS**

| | INTRODUCTION | 72 |
| 61 | HUB BASE | 73 |
| 62 | POLE BRACKET | 74 |
| 63 | T-BRACE | 75 |
| 64 | CHISEL | 76 |
| 65 | HOLD COLLAR | 77 |
| 66 | WEB PULLEY | 78 |
| 67 | SPOKE WHEEL | 79 |
| 68 | VALVE BODY | 80 |
| 69 | GAUGE COVER | 81 |
| 70 | SPINDLE BRACKET | 82 |

**CHAPTER 8 AUXILIARY AND INQUIRY**

| | INTRODUCTION | 83 |
| 71 | TRUE LENGTH OF A LINE | 84 |
| 72 | BEARING OF A LINE | 85 |
| 73 | TRUE SLOPE OF A LINE | 86 |
| 74 | TRUE SIZE AND SHAPE OF A PLANE | 87 |
| 75 | PIERCING POINT BY AUXILIARY VIEW | 88 |
| 76 | SECONDARY AUXILIARY VIEW | 89 |
| 77 | GAUGE BLOCK | 90 |
| 78 | ANGLE PLATE | 91 |
| 79 | ANGLE BRACKET | 92 |
| 80 | STOCK BLOCK | 93 |

**CHAPTER 9 DEVELOPMENTS**

| | INTRODUCTION | 94 |
| 81 | BOX DEVELOPMENT | 95 |
| 82 | CONE DEVELOPMENT | 96 |
| 83 | PYRAMID DEVELOPMENT | 97 |
| 84 | INTERSECTIONS OF TWO LINES | 98 |
| 85 | INTERSECTION OF A LINE AND PLANE | 99 |
| 86 | TRUNCATED CYLINDER DEVELOPMENT | 100 |
| 87 | TRUNCATED HEXAGON | 101 |
| 88 | TWO-PIECE ELBOW DEVELOPMENT | 102 |
| 89 | TRUNCATED CONE | 103 |
| 90 | RECTANGLE TO CIRCLE | 104 |

**CHAPTER 10 3D MODELS**

| | INTRODUCTION | 105 |
| 91 | VISE BLOCK | 106 |
| 92 | T-BLOCK | 107 |
| 93 | BEARING BLOCK | 108 |

| PROBLEM | | PAGE |
|---|---|---|
| | CHAPTER 10 continued | |
| 94 | POST BRACKET | 109 |
| 95 | SPROCKET | 110 |
| 96 | STOP BUTTON | 111 |
| 97 | ROCKER ARM | 112 |
| 98 | SLIDE STOP | 113 |
| 99 | STEP-V PULLEY | 114 |
| 100 | CASTER | 115 |

## CHAPTER 11  SOLID MODELING ADVANCE MODELING EXTENSION (A.M.E.)

| | | |
|---|---|---|
| 101 | INTRODUCTION | 116 |
| 102 | GUIDE PLATE | 117 |
| 103 | TEE STOPS | 118 |
| 104 | SHAFT BLOCK | 119 |
| 105 | COLLAR GUIDE | 120 |
| 106 | GUIDE BLOCK | 121 |
| 107 | SHAFT GUIDE | 122 |
| 108 | LINK UNIT | 123 |
| 109 | HOUSING UNIT | 124 |
| 110 | HOUSING BLOCK | 125 |
| | HAND CLAMP | 126 |

## DRAWING TESTS

| | | |
|---|---|---|
| A-1 | DRAWING TEST | 127 |
| A-2 | EDITING TEST | 128 |
| A-3 | ARRAYING TEST | 129 |
| A-4 | DIMENSIONING TEST | 130 |
| A-5 | DIGITIZING TEST | 131 |
| A-6 | INQUIRY DRAWING TEST | 132 |

## REVIEW QUESTIONS

| | | |
|---|---|---|
| A | REVIEW QUESTIONS A AND ANSWERS | 133-144 |
| B | REVIEW QUESTIONS B AND ANSWERS | 145-155 |
| A-7 | GRAPHIC MATCHING | 156 |
| A-8 | ARRAYING REVIEW | 157 |
| A-9 | MATCH DIM VARIABLES | 158 |

# PREFACE

COMPUTER-AIDED DRAFTING (CAD) IS ONE OF THE MOST RAPIDLY EXPANDING AREAS IN THE INDUSTRIAL SECTOR OF OUR COUNTRY. IN RECENT YEARS, THERE HAS BEEN AN INCREASING DEMAND FOR INDIVIDUALS WITH KNOWLEDGE AND EXPERIENCE OF CAD SYSTEMS AND OPERATIONS. THIS DEMAND WILL BE MORE ACUTE AS THE COST OF BOTH HARDWARE AND SOFTWARE CONTINUES TO BECOME MORE AFFORDABLE TO ANY SIZE OPERATION.

EMPLOYERS ARE DEMANDING THAT PEOPLE EMERGING FROM DRAFTING AND ENGINEERING PROGRAMS HAVE ATTAINED BOTH KNOWLEDGE AND SKILLS OF CAD. THIS TEXT IS DESIGNED TO HELP THE STUDENT ACQUIRE THE NECESSARY SKILLS TO BECOME PROFICIENT IN CAD OPERATION. BECAUSE AUTOCAD® IS THE INDUSTRY STANDARD AND MOST FREQUENTLY ENCOUNTERED CAD PROGRAM, THIS BOOK EMPHASIZES COMMANDS AND OPTIONS FROM THE AUTOCAD® SOFTWARE.

THE TEXT IS ORGANIZED IN 11 CHAPTERS DESIGNED TO BUILD CAD SKILLS. THE DRAWING PROBLEMS ARE TAKEN PRIMARILY FROM THE MECHANICAL AND MANUFACTURING INDUSTRY AND SIMULATE REAL DRAWING PROBLEMS.

CHAPTERS 1 AND 2 ARE DESIGNED TO HELP THE INDIVIDUAL LEARN THE BASIC DRAWING AND EDITING COMMANDS. THE PROBLEMS IN THIS SECTION ARE PROGRESSIVE, MEANING THE PROBLEMS BECOME MORE DIFFICULT AND CHALLENGING TO ALLOW COMPREHENSIVE SKILL DEVELOPMENT.

CHAPTERS 3 AND 4 STRESS THE ADVANCED TOPICS OF BLOCK, ATTRIBUTES, SCALE, FACTORS, AND DIMENSIONING. THESE CHAPTERS BUILD THE SKILLS THAT ALLOW THE OPERATOR TO EFFECTIVELY MANIPULATE THE COMMANDS TO THEIR ADVANTAGE. AFTER COMPLETING THESE FIRST FOUR CHAPTERS, THE INDIVIDUAL SHOULD BE COMFORTABLE WITH THE BASIC MANIPULATION OF THE AUTOCAD® SOFTWARE.

THE BALANCE OF THIS TEXT USES TYPICAL DRAFTING PROBLEMS FROM SEVERAL DISCIPLINES TO PROVIDE PRACTICAL APPLICATIONS TO HELP DEVELOP CAD PROFICIENCY. IN CHAPTERS 5 AND 6 THE INDIVIDUAL PREPARES DRAWINGS FROM THE THREE BASIC ORIENTATIONS: ORTHOGRAPHIC, ISOMETRIC, AND OBLIQUE. THESE EXERCISES PROMOTE NOT ONLY CAD SKILLS BUT ALSO GENERAL DRAFTING SKILLS.

CHAPTER 7 EXPOSES THE INDIVIDUAL TO SECTIONS AND HATCHING, INCLUDING THE COMPLETION OF DRAWING WITH ALL TYPES OF SECTIONS (ie., HALF, FULL, REMOVED, AND REVOLVED.)

THE REMAINING CHAPTERS DEAL WITH MORE ADVANCED TOPICS THAT INCLUDE DESCRIPTIVE GEOMETRY, DEVELOPMENTS, AND THREE-DIMENSIONAL MODELS.

THE APPENDIX INCLUDES SEVERAL DRAWING TESTS THAT CHALLENGE THE INDIVIDUAL TO INCREASE BOTH SPEED AND ACCURACY OF CAD OPERATION. EACH TEST IS DESIGNED TO CHALLENGE SPECIFIC SKILLS AND USE OF COMMANDS DEVELOPED IN THE COURSE OF LEARNING AND WORKING THROUGH THE ASSIGNMENTS. THE REVIEW QUESTIONS ARE INCLUDED TO REINFORCE THE COMMANDS, OPTIONS, AND CONCEPTS OF CAD.

IT IS MY HOPE THAT YOU FIND THIS TEXT BOTH PRODUCTIVE AND CHALLENGING. GOOD LUCK IN YOUR JOURNEY FROM NOVICE TO CAD PROFESSIONAL.

MARK A. SIGL

# ACKNOWLEDGMENTS

I WOULD LIKE TO THANK THE INSTRUCTORS AT HUMBOLDT STATE UNIVERSITY AND CALIFORNIA POLYTECHNIC STATE UNIVERSITY WHO INTRODUCED ME TO COMPUTER-AIDED DRAFTING; DKS ASSOCIATES AND SIGL ASSOCIATES WHO GIVE ME THE OPPORTUNITY TO PERFECT MY CAD SKILLS ON MAJOR PROJECTS; AMERICAN RIVER COLLEGE, GRANT UNIFIED SCHOOL DISTRICT AND ITT TECHNICAL INSTITUTE WHO GAVE ME THE CHANCE TO SHARE MY KNOWLEDGE AND EXPERIENCE WITH OTHERS; DELMAR PUBLISHERS, ESPECIALLY MIKE McDERMOTT AND KEVIN JOHNSON WHO HAD THE PATIENCE AND GUIDANCE TO SEE THIS TEXT FROM CONCEPTION TO FRUITION. FINALLY, I WOULD LIKE TO THANK MY WIFE, TAMMY, WITHOUT HER SUPPORT I WOULD NOT HAVE BEEN ABLE TO PERSEVERE AND SEE THIS TEXT TO ITS COMPLETION.

MARK A. SIGL

# CHAPTER 1
# BASIC DRAWING COMMANDS

THIS CHAPTER COVERS THE BASIC DRAWING COMMANDS NECESSARY TO PERFORM THE FUNDAMENTALS OF CREATING LINE WORK TO PRODUCE DRAWINGS.

PROBLEM                                                           TOPIC

1    CREATE A PROTOTYPE DRAWING THAT HAS THE NECESSARY LAYERS, LINE TYPES, COLORS, AND BORDER OUTLINE.
     THIS PROTOTYPE DRAWING WILL SAVE YOU "SETUP TIME" EVERY TIME IT IS USED.
     ONCE THE BORDER DRAWING HAS BEEN SETUP, IT IS BROUGHT INTO A DRAWING AT THE MAIN MENU OPTION 1 "NEW DRAWING" BY GIVING THE
     NAME OF THE NEW DRAWING FOLLOWED BY AN EQUAL (=), THEN THE NAME OF PROTOTYPE DRAWING "BORDER."   EXAMPLE: PRB1=BORDER.
     A PROTOTYPE DRAWING CAN BE OBTAINED FROM A FLOPPY DRIVE OR SUBDIRECTORY BY GIVING THE PATH BEFORE THE NAME OF THE PROTOTYPE.
     EXAMPLE: PRB1=A:BOR

2    PROBLEM EXPOSES YOU TO DRAWING ON DIFFERENT LAYERS AND WITH DIFFERENT LINE TYPES.  ALSO, YOU USE SNAP, GRID, AND ORTHO
     TO HELP YOU POSITION THE LINE WORK.

3    PROBLEM USES SNAP, ORTHO, AND GRID TO HELP DRAW THE OBECTS.  ALSO, YOU COUNT THE GRID DOTS TO OBTAIN THE LENGTH
     OF THE LINE SEGMENTS.

4    USE ABSOLUTE COORDINATES TO DRAW THE OBJECT TO GAIN AN UNDERSTANDING OF EXACT LOCATION AND POSITIONING.
     THIS METHOD OF COORDINATES IS USED IN SURVEYING, ENGINEERING, AND CAM DRAWINGS.

5    USE RELATIVE COORDINATES TO DRAW THE OBJECT, THE RELATIONSHIP OF RELATING TO THE LAST POINT OF ENTRY.
     THIS METHOD IS USED IN ALL FORMS OF DRAFTING.

6    USE POLAR COORDINATES TO DRAW THE OBJECT, TO DEMONSTRATE THE RELATIONSHIP OF DISTANCE AND ANGLE.
     POLAR COORDINATES ARE USED EXTENSIVELY IN SURVEYING, CIVIL ENGINEERING, AND MECHANICAL DRAWING.
     YOU MUST LEARN ALL THREE METHODS BECAUSE YOU NEED TO USE THEM INTERCHANGEABLY.

7    THIS DRAWING USES CIRCLES, OSNAP, AND BREAK.

8    USE ARCS AND DONUTS TO HELP YOU CREATE CURVES AND SHAPES.

9    DRAWING AIDS YOU IN UNDERSTANDING HOW TO USE TEXT AND DTEXT COMMANDS AND THEIR OPTIONS.  THIS DRAWING ALSO
     BRIEFLY COVERS THE STYLE COMMAND AND HOW TO MODIFY THE BASIC FONTS.
     THIS COMMAND IS USED TO PLACE TEXT ON A DRAWING IN THE FORM OF NOTES.

10   POLYGON COMMAND CAN PRODUCE SHAPES FASTER THAN CREATING THE SHAPE FROM SCRATCH.  THIS COMMAND WITH THE OFFSET COMMAND
     DOUBLES YOUR SPEED.  USE BREAK TO REMOVE LINE SEGMENTS.

PROBLEM 1    BORDER PROTOTYPE DRAWING

THIS BORDER DRAWING IS USED AS A BASE FOR THE WORKING PROBLEMS THAT FOLLOW  AND WILL BE UPDATED AS THE PROBLEMS PROGRESS.
START A NEW DRAWING CALLED "BORB."   THIS  DENOTES A "B" SIZE BORDER.

SET DRAWING LIMITS 17x11

LOAD LINETYPES NEEDED AS SPECIFIED BELOW AND ANY OTHERS YOU MAY WANT
CREATE THE LAYERS SETUP AS LISTED

| LAYER NAMES | COLOR | LINETYPE | LAYER USE |
|---|---|---|---|
| 0 | WHITE | CONTINUOUS | SPECIAL LAYER FOR BLOCKS |
| BORDER | WHITE | CONTINUOUS | BORDER LINE |
| OBJ | BLUE | CONTINUOUS | OBJECT LINES |
| CON | CYAN | CONTINUOUS | CONSTRUCTION LINES |
| HL | GREEN | HIDDEN | HIDDEN LINES |
| CL | YELLOW | CENTER | CENTER LINES |
| PL | MAGENTA | PHANTOM | PHANTOM LINES |
| TEXT | RED | CONTINUOUS | TEXT |
| DIM | RED | CONTINUOUS | DIMENSIONING |

PLOTTING PEN SETUP INFORMATION

| COLORS | PLOTTING PEN WIDTHS |
|---|---|
| RED | .35 |
| YELLOW | .35 |
| GREEN | .35 |
| CYAN | .35 |
| BLUE | .50 |
| MAGENTA | .35 |
| WHITE | .50 |

(OTHER PEN SETUPS CAN BE USED,
USE DIFFERENT PEN THICKNESSES OR
COLORS FOR LINE WEIGHTS)

SET LAYER TO OBJ
SET GRID AND SNAP TO .25"
SET VIEWRES TO 5000
SET BLIPMODE TO "OFF" UNLESS YOU LIKE TO SEE BLIPS (+) EVERY TIME YOU PICK
SET REGENAUTO TO "ON" TO CONTINUOUSLY UPDATE THE DRAWING DATABASE
SET UCSICON TO "OFF"
SET UNITS TO DECIMAL AND TO TWO PLACE VALUE

SAVE "BORB" TO YOUR FLOPPY AND END YOUR PROTOTYPE DRAWING.
NOW ARE YOU READY TO START WORKING ON THE PROBLEMS TO INCREASE YOUR PROFICIENCY.
BE SURE TO SAVE YOUR DRAWINGS TO YOUR FLOPPY SO YOU HAVE A COPY AND KEEP A HARD COPY FOR YOUR PORTFOLIO.


TO USE THE PROTOTYPE "BORB" DRAWING, YOU NEED TO CALL UP THE DRAWING WHEN STARTING A NEW DRAWING ON THE MAIN MENU BY
ISSUING AN EQUAL (=) AFTER THE DRAWING NAME GIVEN.  EXAMPLE: PRB1=BORB.  A PROTOTYPE DRAWING CAN BE OBTAINED FROM A FLOPPY DRIVE
OR SUBDIRECTORY BY GIVING THE PATH BEFORE THE NAME OF THE PROTOTYPE.  EXAMPLE: PRB1=B:\DWG\BORB.  THIS WILL TAKE THE "BORB" DRAWING
OFF THE "A" DRIVE AND OUT OF THE SUBDIRECTORY CALLED "DWG."


NOTE: THIS PROTOTYPE SAVES YOU A LOT OF SETUP TIME.  THE DRAWING STAYS THE WAY IT HAS BEEN SETUP THUS ELIMINATING THE NEED TO
REDO THE SET UP EACH TIME. OTHER LAYERS AND LINETYPES CAN BE ADDED AS NEEDED. IT IS GOOD PRACTICE TO HAVE A PROTOTYPE BASE
DRAWING AS DIVERSE AND SIMPLE AS POSSIBLE.  OVERLOADING A DRAWING WITH COMPONENTS NOT USED MOST OF THE TIME IS A WASTE
i.e., A BORDER DRAWING WITH GREATER THAN 100 LAYERS, LINETYPES, AND BLOCKS.

PROBLEM 1 (CONTINUED)
BORDER PROTOTYPE
DRAWING EXAMPLE

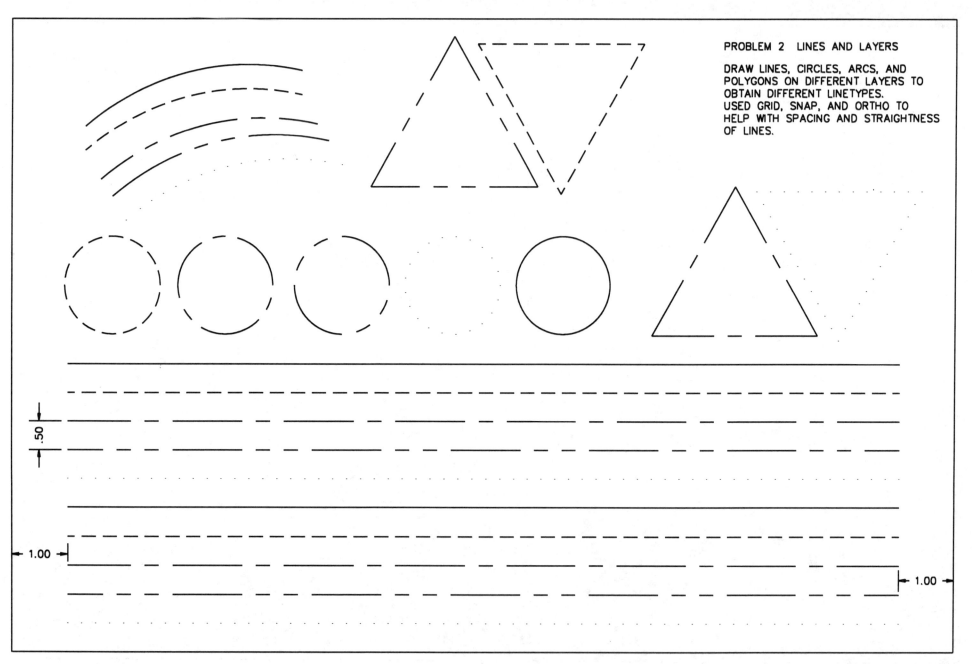

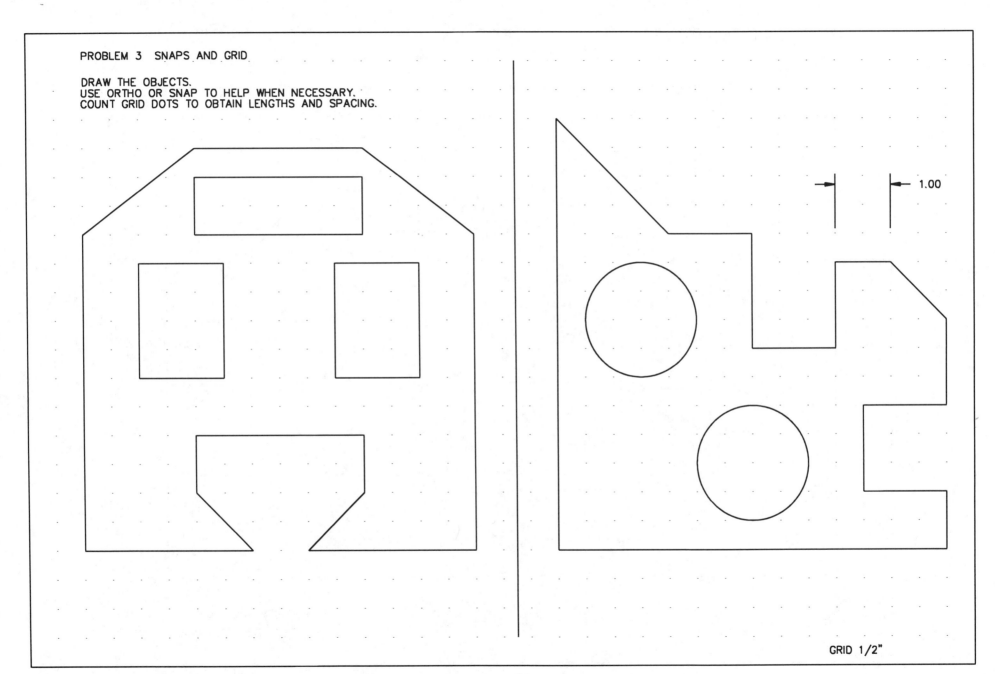

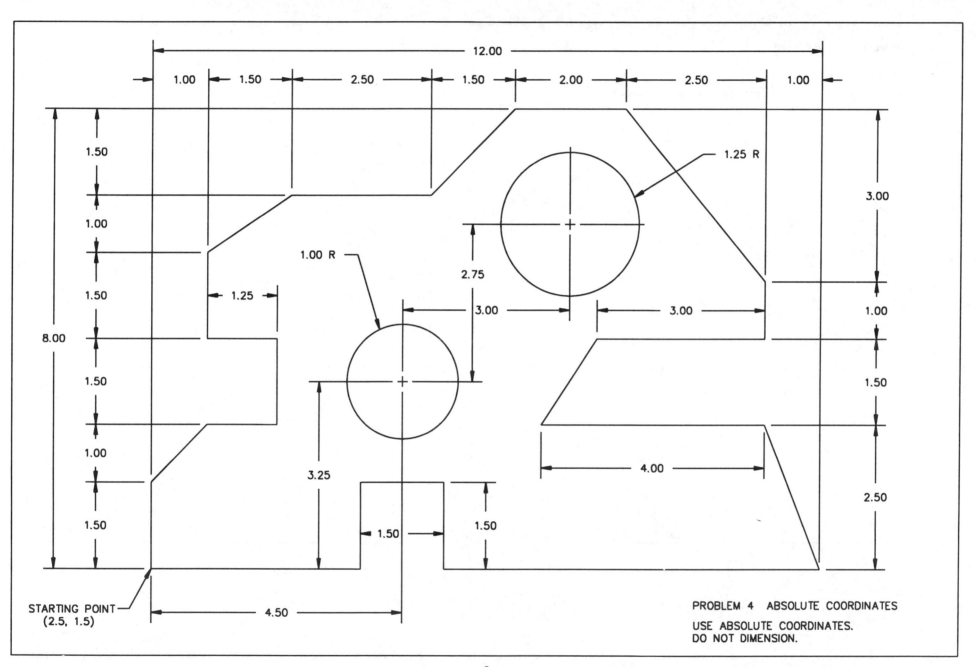

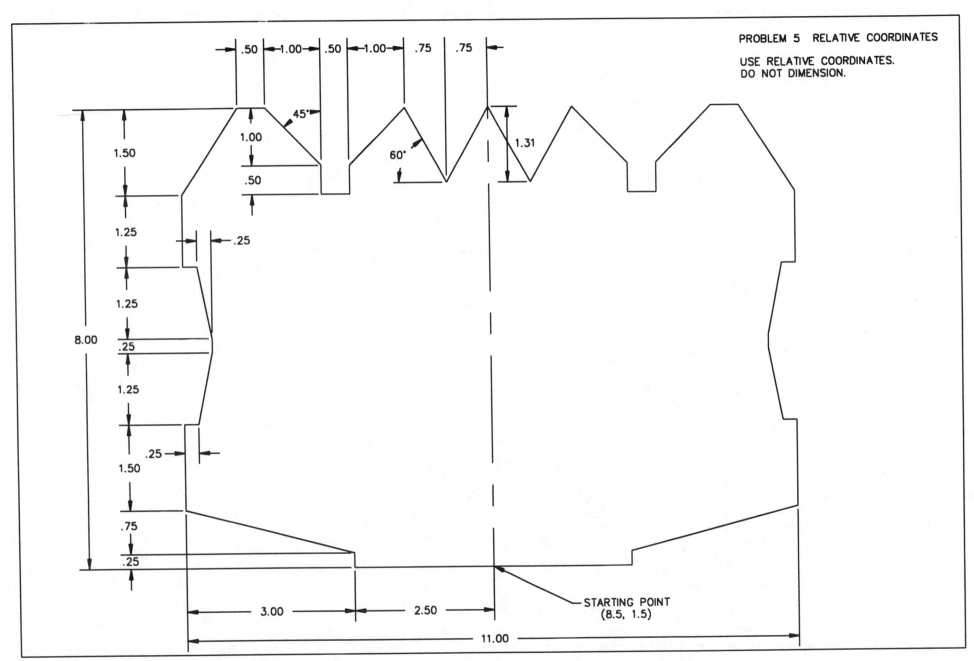

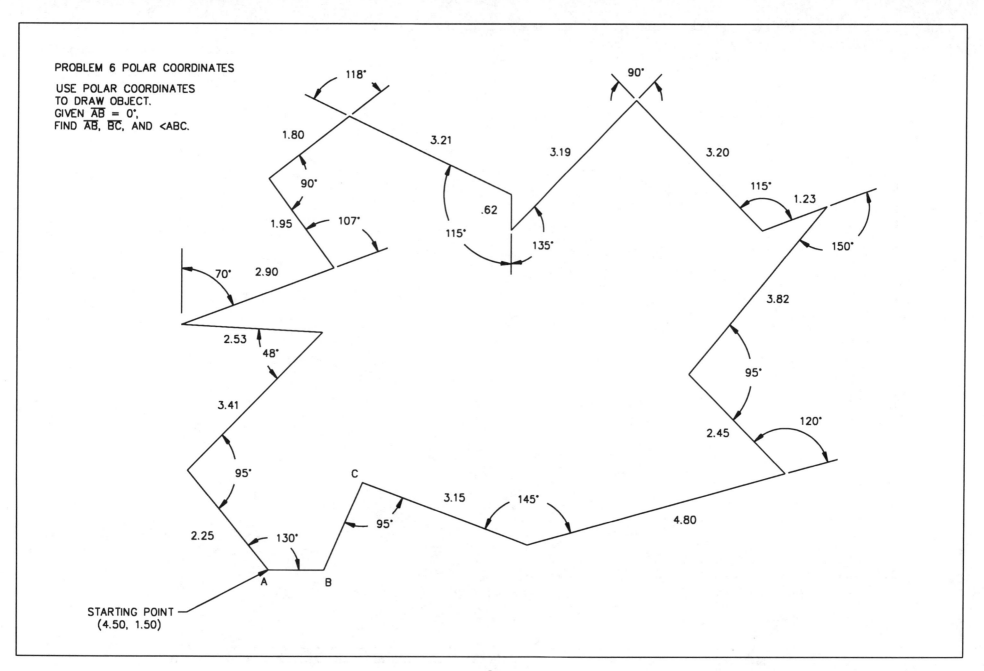

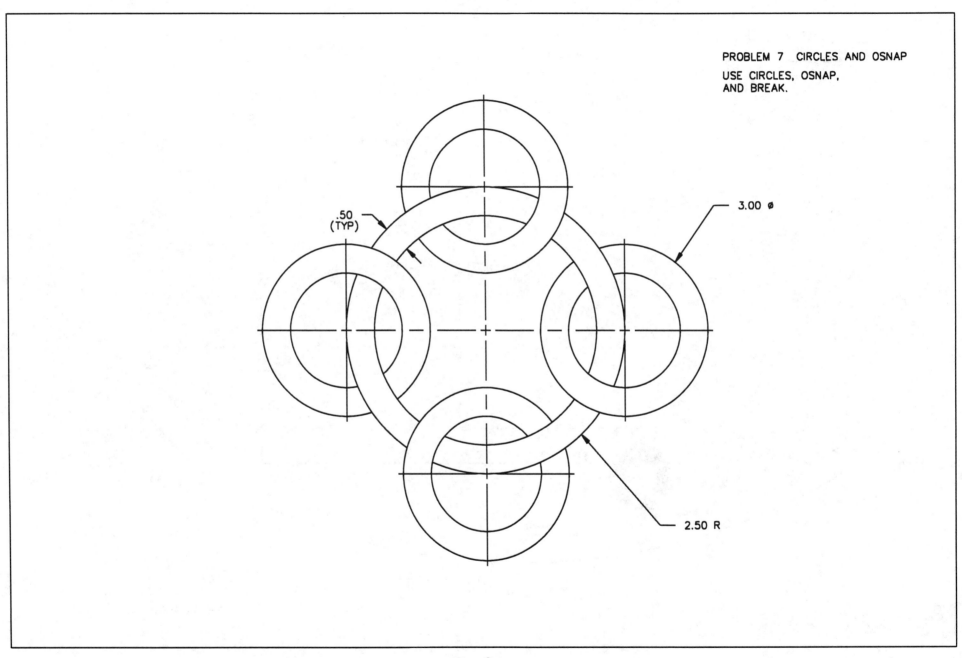

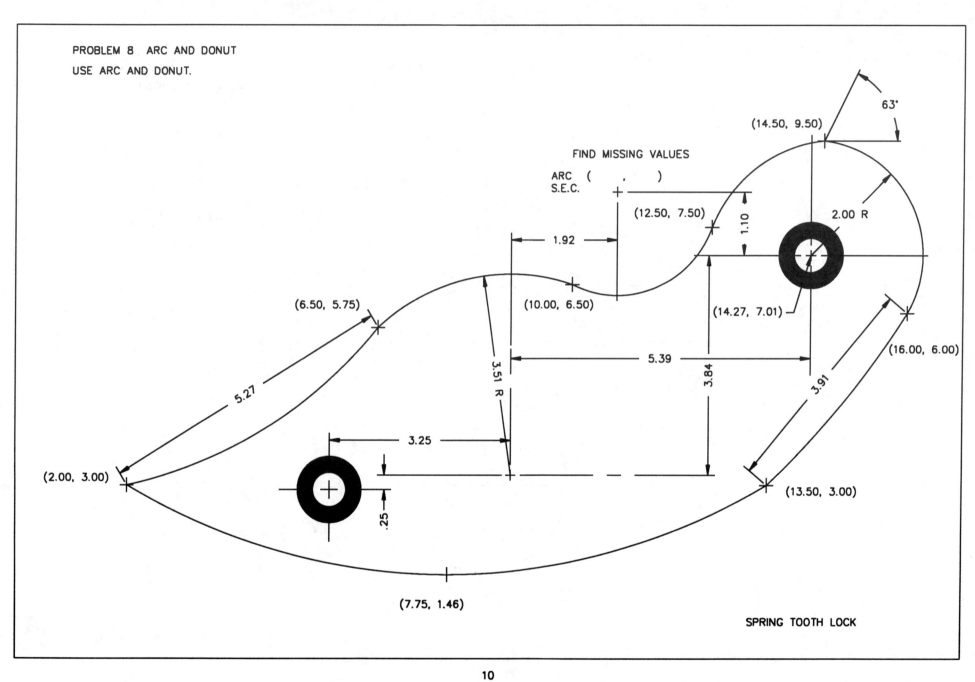

PROBLEM 9   DTEXT, TEXT, AND STYLE


USE STYLE COMMAND TO CREATE FIVE DIFFERENT TEXT STYLES
STYLE NAME:  A NAME TO RECALL THE TEXT MODIFIED FILE (UP TO 31 CHARACTERS)
FONT FILE: NAME OF ASSOCIATED FONT FILE
HEIGHT:   A NUMBER FOR FIXED HEIGHT (USE 0 HEIGHT TO PROMPT FOR HEIGHT EACH TIME)
WIDTH FACTOR:   EXPANSION / COMPRESSION FACTOR
OBLIQUING ANGLE: SLANT ANGLE (0 NORMAL, + FORWARD, – BACKWARDS)
BACKWARDS:   YES/NO SDRAWKCAB SDROW SETIRW
UPSIDE-DOWN:  YES/NO FLIPS WORDS OVER
VERTICAL:     YES/NO ORIENTATION VERTICAL/HORIZONTAL

| STYLE NAME: | STANDARD | STANDARD2 | STANDARD3 | STANDARD4 | VERTICAL |
|---|---|---|---|---|---|
| FONT FILE: | SIMPLEX | ROMANC | ITALIC | GOTHICE | ROMANS |
| HEIGHT: | 0 | 0 | .25 | 0 | 0 |
| WIDTH FACTOR: | 1.00 | 1.2 | .8 | 1 | 1 |
| OBLIQUING ANGLE: | 0 | 10 | 15 | −10 | 0 |
| BACKWARDS: | NO | N | N | N | N |
| UPSIDE-DOWN: | NO | N | N | Y | N |
| VERTICAL: | NO | N | N | N | Y |


USING TEXT AND DTEXT ALLOWS YOU TO PLACE TEXT A DRAWING USING ONE OF THE JUSTIFICATIONS AVAILABLE.
THE DIFFERENCE BETWEEN TEXT AND DTEXT IS THAT DTEXT (DYNAMIC TEXT) LETS YOU SEE THE LETTERS AS THEY ARE BEING TYPED.
THIS ALLOWS YOU TO SEE THE PLACEMENT AND SPACING OF TEXT ON YOUR DRAWING.  YOU NEED TO REMEMBER TO USE THE "RETURN KEY"
TWICE TO FIX THE TEXT TO THE DRAWING, OTHERWISE TEXT DISAPPEARS.
START POINT:  BEGINS TEXT, LEFT JUSTIFIED, AT THE POINT PICKED.
ALIGN:        PROMPTS FOR TWO POINTS, VARYING THE CHARACTER HEIGHT BASED ON NUMBER OF CHARACTERS TO FIT TEXT BETWEEN SPECIFIED POINTS,
CENTER:       PROMPTS FOR A POINT AND THEN CENTERS TEXT BASELINE ON THAT POINT.
FIT:          WORKS LIKE ALGIN BUT USES A SPECIFIED HEIGHT AND ADJUSTS THE CHARACTERS WIDTH SO TEXT FITS BETWEEN THE TWO POINTS.
MIDDLE:       PROMPTS FOR A MIDDLE POINT AND CENTERS TEXT BOTH VERTICALLY AND HORIZONTALLY.
RIGHT:        PROMPTS FOR A POINT, THEN RIGHT JUSTIFIES TEXT TO THAT POINT.
STYLE:        ALLOWS YOU TO "SWITCH" TO ANOTHER TEXT STYLE THAT YOU HAVE CREATED.  STYLE PROMPT IS FOR THE START POINT PROMPT.
NULL REPLY:  CAUSES THE NEW TEXT STRING TO START DIRECTLY BELOW THE LAST STRING USING THE LAST JUSTIFICATION OPTION CHOSEN.

USE STANDARD 2 WITH .25 HEIGHT & STARTING POINT

*USE STANDARD 3 WITH THE RIGHT JUSTIFICATION*

*USE STANDARD 3 WITH MIDDLE JUSTIFICATION*
*NOTICE THAT THERE IS NO TEXT HEIGHT PROMPT*

USE STANDARD 2 WITH ALIGN JUSTICATION

USE STANDARD 1 WITH THE FIT OPTION AND .20 HEIGHT

SEE WHAT HAPPENS TO FIT WHEN PICK ORDER IS REVERSED

USE STANDARD 3 WITH ALIGN JUSTIFICATION AND PICK IN REVERSE ORDER

USE STANDARD 3 WITH FIT

PROBLEM 9 CONTINUED
DTEXT, TEXT, AND STYLE

USE TEXT AND. DTEXT COMMAND.
USE STYLE COMMAND TO MODIFY TEXT FILES.
USE STYLE OPTION IN TEXT AND DTEXT TO SWITCH TEXT FILES.
EXERCISE GIVES YOU EXPERIENCE WITH
THESE COMMANDS AND OPTIONS.

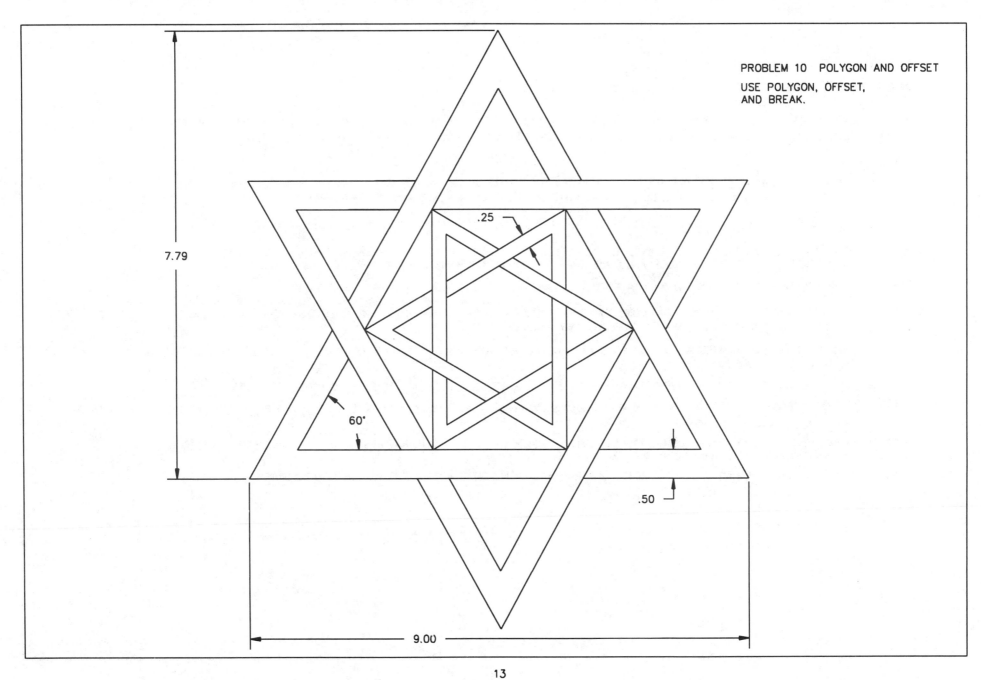

# CHAPTER 2
# BASIC EDITING COMMANDS

THIS CHAPTER COVERS THE BASIC CONCEPTS OF EDITING ENTITIES IN A DRAWING. THROUGHLY UNDERSTANDING HOW ENTITIES ARE EDITED OR MODIFIED TO PRODUCE A DRAWING MAKES YOUR SKILLS AS CAD DRAFTER MORE VALUABLE AND INCREASES YOUR SPEED.

PROBLEM                                                                          TOPIC

11      RECTANGULAR ARRAY SAVES YOU MUCH TIME WHEN LAYING OUT ROWS AND COLUMNS OF OBJECTS IN A DRAWING.
        AN EXAMPLE IS BUILDING COLUMNS, DRILL HOLES, NUTS OR BOLTS ON A PLATE AND GRAPHIC ARTS.

12      POLAR ARRAY IS USED TO COPY ITEMS SUCH AS HOLES, NUTS BOLTS, ELECTRONICS, AND GEARS IN A CIRCULAR FASHION.

13      TRIM COMMAND REMOVES LINE SEGMENTS THAT CROSS ANOTHER LINE OR THAT ARE BETWEEN TWO LINE SEGMENTS.
        THIS COMMAND IS FASTER AND MORE ACCURATE THAN THE BREAK COMMAND IN REMOVING A LINE SEGMENT.

14      COPY COMMAND ALLOWS YOU TO MAKE A REPLICA OF AN ENTITY AND PLACE IT AT ANY LOCATION.

15      CHANGE COMMAND ALLOWS YOU TO MODIFY EXISTING ENTITIES AND THEIR PROPERTIES; INCLUDES ITEMS SUCH AS LAYER, LINETYPE, COLOR, LENGTH, THICKNESS, ELEVATION, HEIGHT, INSERTION POINT, AND MANY OTHER THINGS.

16      CHAMFER COMMAND CREATES A TAPERED EDGE AT A CORNER INTERSECTION. COMMAND TRIMS OR LENGTHENS THE LINE TO FORM
        THE DISTANCES SPECIFIED.

17      FILLET ROUNDS THE CORNER TO THE RADIUS SPECIFIED. IT ALSO LENGHTENS OR TRIMS THE CORNER AS NEEDED.

18      MIRROR COMMAND COPIES AND ROTATES THE OBJECT AROUND AN AXIS. THIS COMMAND IS EXCELLENT FOR PRODUCING THE OTHER IDENTICAL
        HALF OR QUARTER OF AN OBJECT.

19      STRETCH, PLINE, AND PEDIT COMMANDS ARE USED TO SHORTEN OR ELONGATE ENTITIES. THE PLINE AND PEDIT CREATE
        AND MODIFY LINES WITH VARYING THICKNESS.

20      SCALE COMMAND ALLOWS YOU TO INCREASE OR DECREASE ENTITY SIZES BY ANY FACTOR OR RELATIONSHIP.

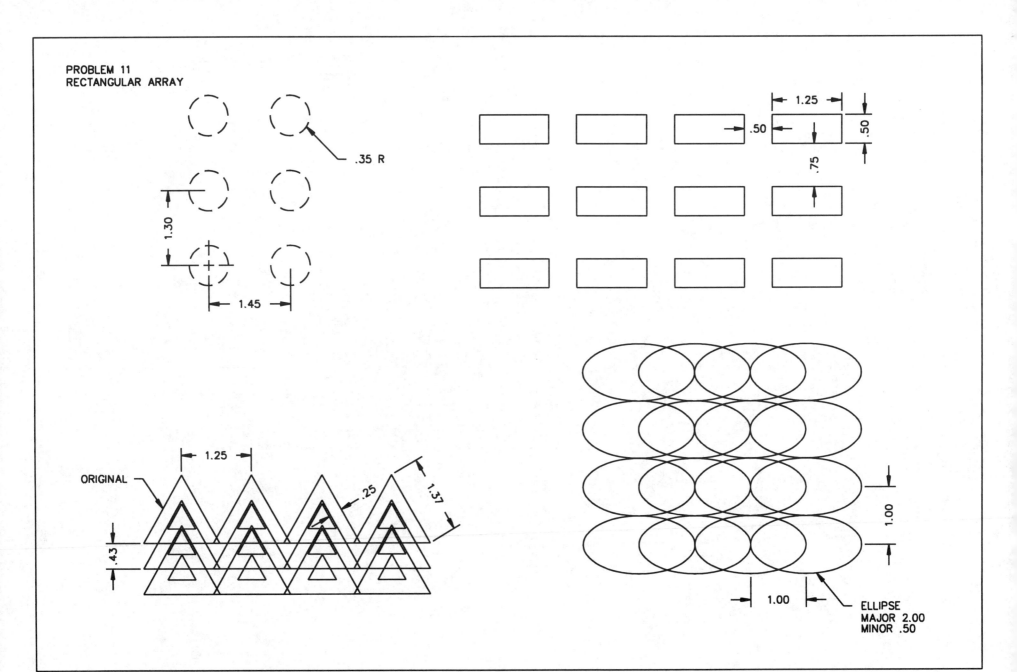

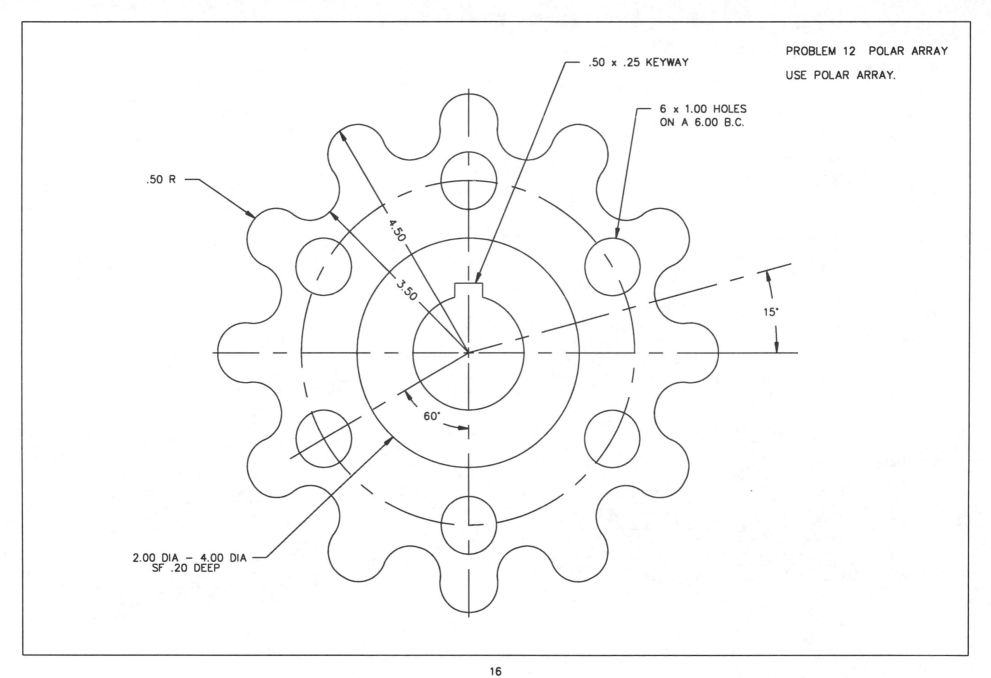

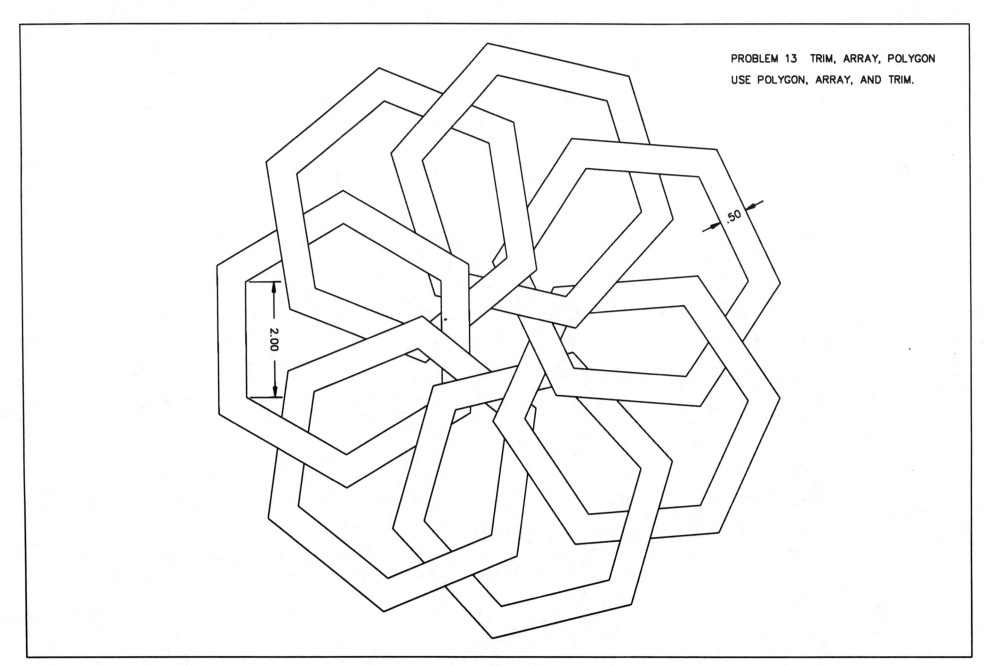

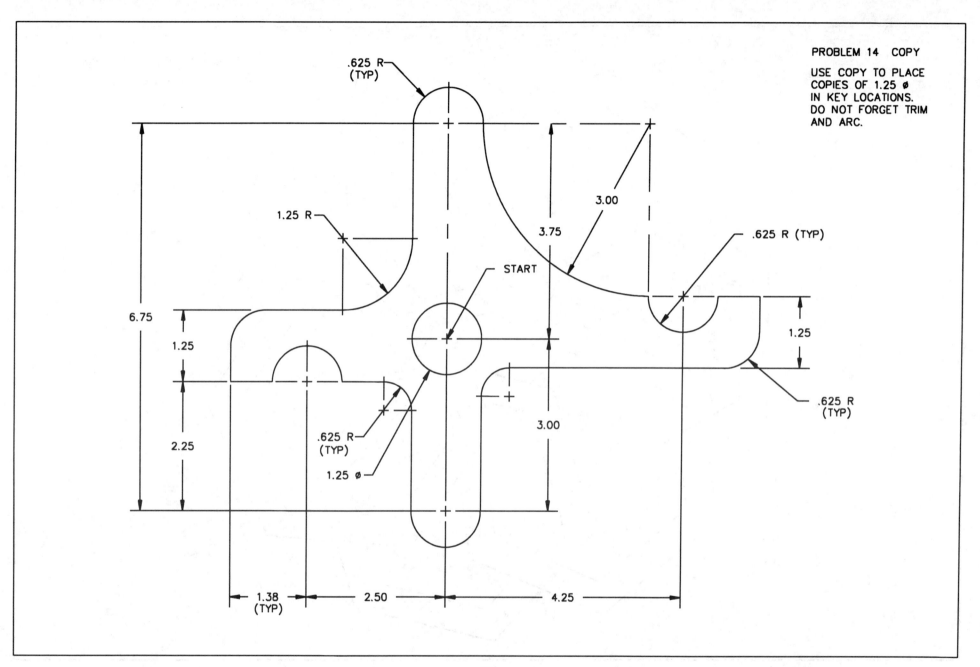

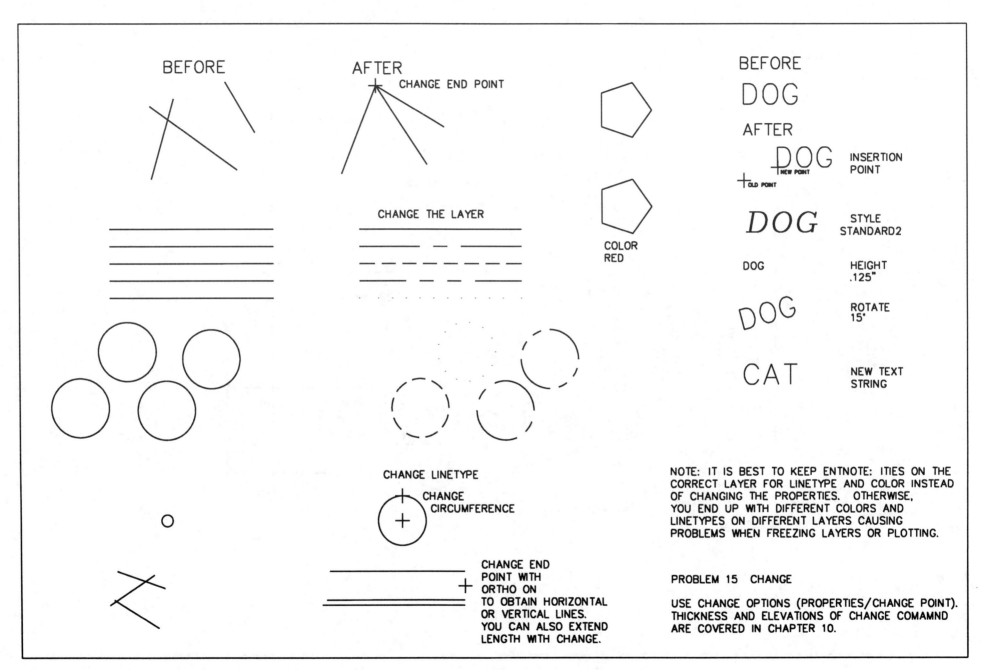

PROBLEM 16  CHAMFER
USE CHAMFER.

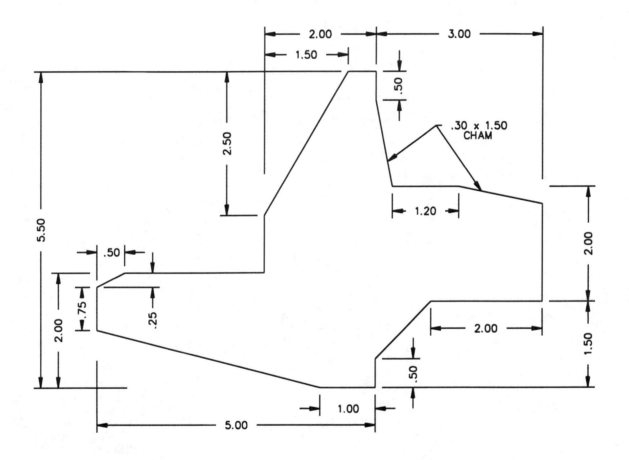

PROBLEM 17   FILLET

USE FILLET TO
ROUND EDGES TO
APPROPRIATE RADIUS.
NOTE: FILLETS ARE .30 R.

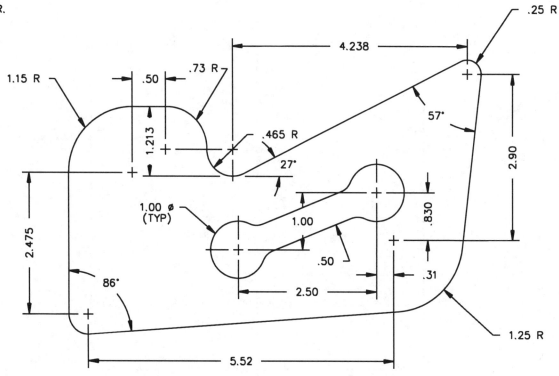

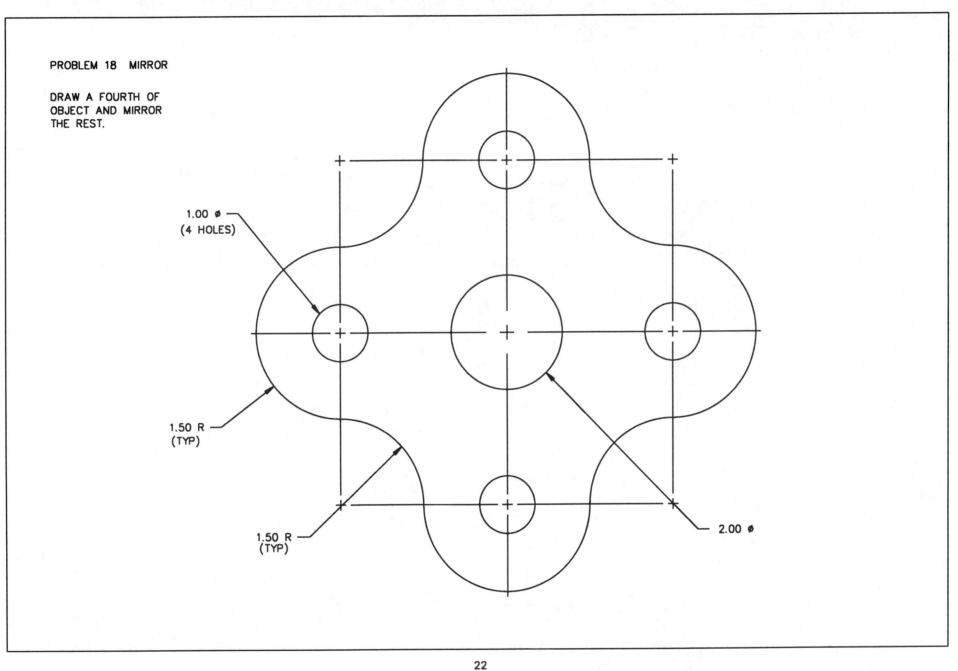

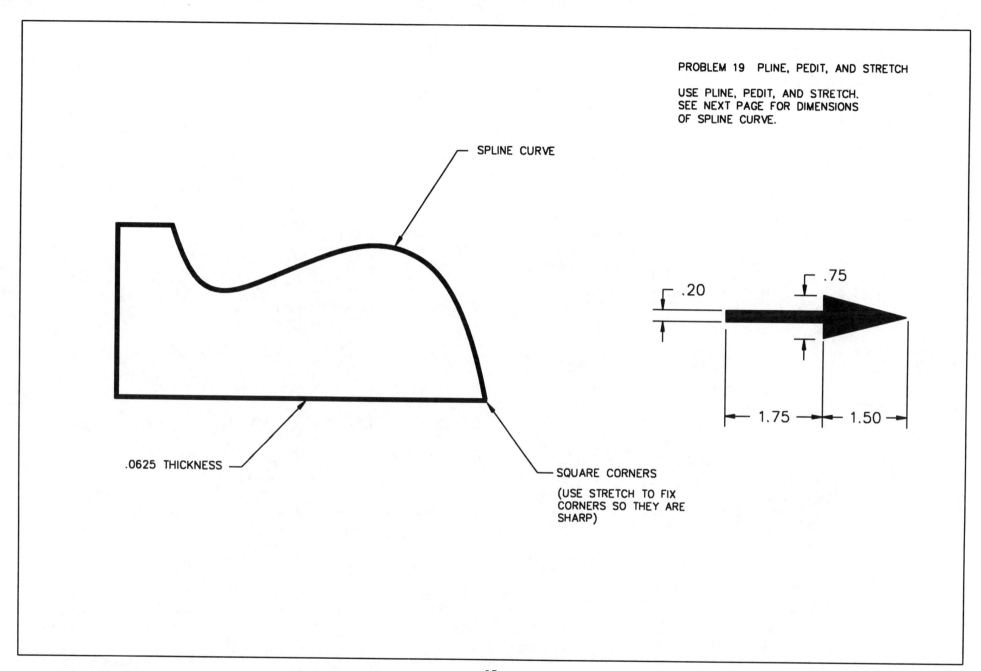

PROBLEM 19 CONTINUED

DIMENSIONS FOR
PLINE OBJECT.

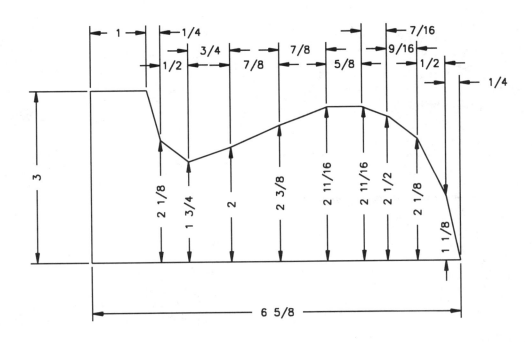

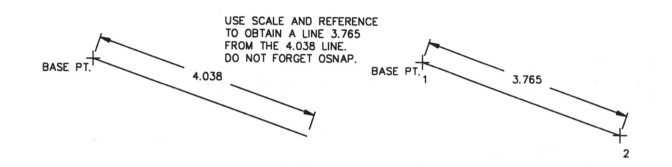

PROBLEM 20  SCALE AND ROTATE

USE SCALE AND ROTATE.

USE SCALE AND REFERENCE TO OBTAIN A LINE 3.765 FROM THE 4.038 LINE. DO NOT FORGET OSNAP.

1/8 TEXT SIZE SCALED TO THREE TIMES

# 1/8 TEXT SIZE SCALED TO THREE TIMES

USE 1.50 LINE AND SCALE WITH DIFFERENT BASE POINT LOCATIONS. SCALE EACH BY A FACTOR OF 1.5.

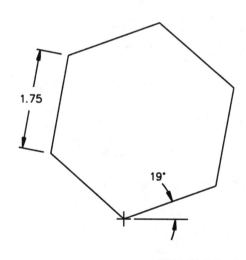

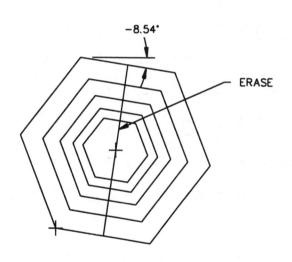

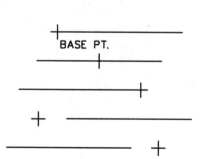

ROTATE POLYGON BACK TO -8.54°.  MAKE SECOND COPY AND SCALE 76.6% OF ORIGINAL SIZE. SCALE BASE IS CENTER.  DO FOUR TIMES.

# CHAPTER 3
# BLOCKS, ATTRIBUTES, & SCALE FACTORS

THIS CHAPTER COVERS CREATING BLOCKS, ATTRIBUTES, AND SCALE FACTORS.  BY UNDERSTANDING THE FUNDAMENTALS OF BLOCKS, ATTRIBUTES, AND SCALE FACTORS, YOU SAVE VALUABLE DRAWING TIME.

PROBLEM                                                                          TOPIC

21    CREATING A BLOCK LIBRARY HAS SEVERAL ADVANTAGES SUCH AS ELIMINATING MAKING THE SAME SYMBOLS MORE THAN ONCE, UNIFORMITY OF SYMBOLS, EASIER UPDATING OF SYMBOLS, AND ALLOWS MORE TIME FOR DESIGNING RATHER THAN REDRAWING.

22    ASSEMBLY OF A DRAWING USING A BLOCK LIBRARY.  BE SURE TO NOTE WAYS TO IMPROVE YOUR BLOCK'S CHARACTISITICS TO INCREASE SPEED, EASE OF INSERTING, AND BLOCK MODIFICATIONS.

23    CREATING A BORDER DRAWING WITH ATTRIBUTES TO TITLE BOX.  THIS SHOWS YOU HOW INFORMATION CAN BE INCORPORATED INTO A PROTOTYPE DRAWING.  BE SURE TO PLAN CAREFULLY IN SETTING UP THIS PROTOTYPE SO YOU DO NOT HAVE TO KEEP UPDATING THE DRAWING CONSTANTLY.

24    EXERCISE SHOWS YOU HOW BLOCKS CAN BE ASSEMBLED TO CREATE A NEW BLOCK.  THEN ONE OF THE ORIGINAL BLOCKS CAN BE UPDATED, THEREBY CHANGING THE NEW BLOCK GROUPING.

25    USING A BLOCK WITH ATTRIBUTES ALLOWS YOU TO SPEED UP SUCH TASKS AS PARTS LISTS, SCHEDULES, NOTES, KEYS, AND SYMBOLS. THE USE OF ATTRIBUTES ALSO ALLOWS YOU TO EXTRACT THE INFORMATION TO BE USED IN WORD PROCESSING AND SPREADSHEETS.

26    A "C" SIZE BORDER IS CREATED TO BE USED AT DIFFERENT SCALES AND ALLOWS YOU TO PLOT DRAWINGS AT DIFFERENT SIZES.

27    CREATE A "D" SIZE BORDER FOR YOUR NEEDS OR FOR COMMERICAL NEEDS.  IN ADDITION YOU SHOULD GET USE TO MAKING AND CUSTOMIZING BORDERS FOR YOUR CUSTOMERS.

28    NOW YOU ARE READY TO START PRODUCING DRAWINGS AT DIFFERENT SCALES.  ACTUALLY THE DRAWINGS ARE REAL SIZE, IT'S THE TEXT AND BORDER THAT ARE SCALED.  BE SURE YOU UNDERSTAND SCALING AND SCALE FACTORS BECAUSE YOU HAVE DIFFERENT SCALES ON ONE DRAWING, DETAILS, SCHEDULES, AND NOTES.  DRAWING SCALE IS HALF SIZE.

29    THAT'S RIGHT!  YOU GUESSED IT!  ANOTHER DRAWING WITH A DIFFERENT SCALE.  HOW ELSE ARE YOU GOING TO GET PROFICIENT! DRAWING IS 1/4" = 1'.

30    TO MAKE SURE YOU GET EXPOSED TO ALL TYPES OF SCALES, PROBLEM IS ANOTHER DRAWING WITH A DIFFERENT SCALE OF 1"=20'.

PROBLEM 21 ELECTRICAL SYMBOL LIBRARY

DIRECTIONS

DRAW THE SYMBOLS ON "OBJ" LAYER.
TO OBTAIN ACTUAL SIZE, LOOK IN A DRAFTING,
ELECTRICAL, OR STANDARD BOOK.
PLACE TEXT ON "TEXT" LAYER.
USE OSNAP TO PICK BASE POINT OF BLOCK.
BLOCK SYMBOLS ONLY. USE ABBREVIATIONS AS BLOCK
NAME. DO NOT BLOCK TEXT ALONG WITH SYMBOLS.
REINSERT BLOCK NEXT TO TEXT IN THE SAME DRAWING.
WHEN YOU NEED TO USE LIBRARY SYMBOLS, INSERT
THE LIBRARY INTO YOUR DRAWING. WHEN IT ASKS FOR
INSERTION POINT, PRESS CTRL-C. NOW YOU CAN INSERT
ANY ONE OF THE INDIVIDUAL SYMBOLS USING THE
ABBREVIATED NAME.
THIS LIBRARY METHOD CAN BE USED FOR ANY GROUPING
OF SYMBOLS (ARCHITECTUAL, ELECTRICAL, PIPING, CIVIL).
YOU CAN PURGE EXCESS SYMBOLS WHEN YOU'RE DONE
WITH THE DRAWING. THIS DECREASES THE DRAWING
FILE SIZE FOR STORAGE.

NOTE:
x = INSERTION POINT

PROBLEM 22
ELECTRICAL SCHEMATIC

DRAW THE SCHEMATIC USING
THE ELECTRICAL LIBRARY YOU
HAVE CREATED.

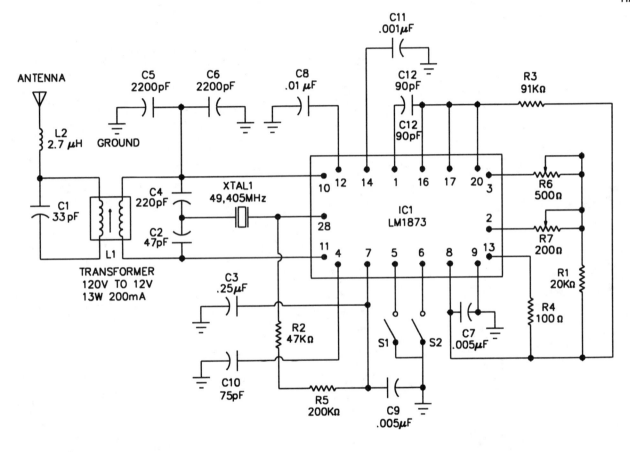

PROBLEM 23   TITLE BOX ATTRIBUTE AND BORDER UPDATE

THIS DRAWING USES A BLOCK CALLED TITLE.  THE BLOCK CONTAINS THE ATTRIBUTE INFORMATION: YOUR NAME, TITLE OF PROBLEM, SCALE OF
DRAWING, AND DATE YOU CREATED DRAWING.  THIS TEXT INFORMATION IS KEPT ON A LAYER CALLED "BORTEXT."  THIS ALLOWS
YOU THE OPTION TO FREEZE THE LAYER FOR ROUGH DRAFTS AND TO SPEED UP REGENS BY REMOVING THE TEXT ON THE BORDER DRAWING.
THE "BORTEXT" LAYER IS RED.  THIS ENSURES THAT THE SAME PEN THICKNESS IS CONSISTENT WITH THE OTHER TEXT ON THE DRAWING.
OF COURSE WE CAN CHANGE THE COLOR ANY TIME BY CHANGING THAT LAYER'S COLOR.

TO START AN ATTRIBUTE, USE "ATTDEF" COMMAND.  YOU HAVE MODES SUCH AS INVISIBLE, CONSTANT, VERIFY, AND PRESET.  THESE MODES ARE USED
TO HELP YOU MODIFY HOW THE INFORMATION IS CONTROLLED.   INVISIBLE: HIDES INFORMATION, UNLESS FORCED.
                                                        CONSTANT: SETS INFORMATION TO A FIXED VALUE.
                                                        VERIFY: IT ASKS IF INFORMATION IS CORRECT.
                                                        PRESET: SETS INFORMATION TO DEFAULT BUT CAN STILL BE EDITED.

ATTDEF
ATTRIBUTE MODES:
ATTRIBUTE TAG:           NAME                    TITLE                           SCALE                   DATE
ATTRIBUTE PROMPT:        ENTER YOUR NAME         ENTER DRAWING TITLE             ENTER DRAWING SCALE     ENTER TODAY'S DATE
DEFAULT VALUE:           "YOUR NAME"             BORDER "B" SIZE                 FULL                    1/1/91
START POINT:             PICK L.L.C.             USE "CENTER" & SELECT CENTER    PICK L.L.C.             PICK L.L.C.
HEIGHT <.125>:           .125                    <CR>                            <CR>                    <CR>
ROTATION ANGLE <0>:      <CR>                    <CR>                            <CR>                    <CR>

BLOCK ATTRIBUTES
BLOCK NAME: TITLE     THIS RECALLS THE ATTRIBUTE FOR YOUR TITLE BOX INFORMATION.
INSERTION POINT: 0,0     SAME AS BORDER ORIGIN
SELECT:   PICK IN ORDER (TITLE, NAME, SCALE, DATE)
          THIS CAUSES THE ATTRIBUTES TO ASK FOR INFORMATION IN THAT ORDER.

AFTER ALL UPDATES HAVE BEEN PERFORMED, SAVE NEW BORDER AS "BORB"

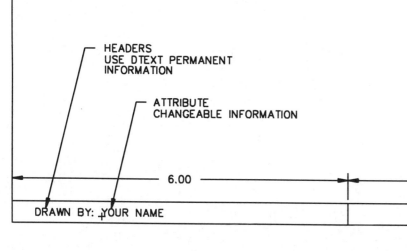
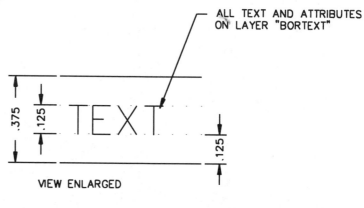

PROBLEM 23  BORDER B

| DRAWN BY: | | SCALE: | DATE: |

STEP 1

CREATE BLOCKS.
BLOCK NAMES.

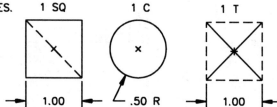

STEP 2

INSERT ALL THREE BLOCKS
WITH SCALE FACTOR 2.

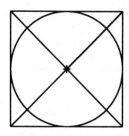

REBLOCK GIVING IT
A NEW NAME "POST"

PROBLEM 24
NESTED BLOCKS

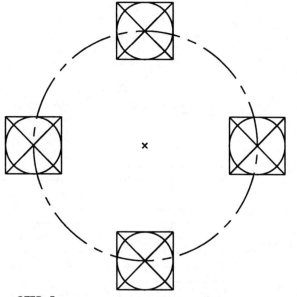

STEP 3

INSERT POST ONTO A 4" ⌀ CIRCLE. USE OSNAP QUAD
WITH A SCALE FACTOR OF .5.
NOW BLOCK THIS AND CALL IT "PLATE."

STEP 4

NOW WE ARE GOING TO REDEFINE SOME OF THE NESTED BLOCKS.
WHAT DO YOU THINK WILL HAPPEN?

BLOCK NAMES

| DRAWN BY: USING AUTOCAD | NESTED BLOCKS | SCALE: FULL | DATE: 6/1/91 |

| | | | | |
|---|---|---|---|---|
| HEADER ⟶ | NO. | PART | REQ | MATL |
| VALUE WITH (−) ⟶ | NO. | PART | REQ | MATL |
| VALUE WITH (spacebar) ⟶ | NO. | PART | REQ | MATL |
| | NO. | PART | REQ | MATL |
| | NO. | PART | REQ | MATL |
| | NO. | PART | REQ | MATL |
| | NO. | PART | REQ | MATL |
| | NO. | PART | REQ | MATL |
| | NO. | PART | REQ | MATL |
| | NO. | PART | REQ | MATL |
| | NO. | PART | REQ | MATL |

(BEFORE BLOCK)

**PROBLEM 25**
**PARTS LIST WITH ATTRIBUTES**

PARTS LISTS ARE FORMED IN MANY WAYS.  THIS EXERCISE ILLUSTRATES FOUR DIFFERENT METHODS.
THE HEADER CAN BE A LABEL OR AN ATTRIBUTE EXPLODED.
USE "ATTDIA" (FOUND IN SETVAR) TO HAVE THE ATTRIBUTES COME UP AS A DIALOGUE BOX WHEN INSERTED.
IF YOU NEED TO SPEED UP REGEN TIME ON DRAWINGS WITH A LOT OF ATTRIBUTES, YOU CAN SET "ATTDISP"
TO "OFF" TO MAKE THEM INVISIBLE UNTIL YOU ARE DONE.  DO NOT FORGET TO SET ATTDISP BACK TO NORMAL.
OR IF YOU HAVE INVISIBLE ATTRIBUTES AND YOU WANT TO SEE THEM, CHANGE ATTDISP TO "ON" AND IT FORCES
ALL ATTRIBUTES TO BECOME VISIBLE.

USE THE PRESET ON ATTRIBUTE MODES.  THIS DEFAULTS TO YOUR VALUE IMMEDIATELY.  YOU CAN THEN
GO BACK USING "DDATTE" OR "ATTEDIT" TO UPDATE THE VALUES AS NEEDED, WHICH LEAVES THE
REST OF THE ATTRIBUTES BLANK. (SEE OPTION 2.)
INSERTING BASE POINT FOR PART 1 AND PART 2 IS 0,0 OR CORNER.
FOR PART 3 AND PART 4 SEE EXAMPLE FOR INSERTING BASE POINT.
SAVE THE PARTS LIST AS (PART1, PART2,...). DO NOT FORGET TO WBLOCK EACH PART LIST.

**OPTION 1**

PARTS LIST CAN BE IN UPPER RIGHT
CORNER, ALLOWING YOU TO FILL FROM
TOP TO  BOTTOM.

| TAG: | PROMPT: | VALUE: | TEXT TYPE: | TEXT LOCATION: | HEIGHT: | ANGLE: |
|---|---|---|---|---|---|---|
| NO. | ENTER ID NUMBER | − or (spacebar) | CENTER | PICK LOCATION | .125 | O |
| PART | ENTER PART NAME | − or (spacebar) | CENTER | PICK LOCATION | .125 | O |
| REQ | ENTER NUMBER REQUIRED | − or (spacebar) | CENTER | PICK LOCATION | .125 | O |
| MATL | ENTER MATERIAL TYPE | − or (spacebar) | CENTER | PICK LOCATION | .125 | O |

**OPTION 3**

OPTION 3  ALLOWS YOU TO BUILD PARTS
LIST AS BIG AS YOU NEED BY INSERTING OR
ARRAYING AS MANY PARTS AS YOU NEED.  TO GET
THE HEADER, EXPLODE MOST UPPER
ATTRIBUTE OR ADD HEADER WITH DDATTE.

**OPTION 2**

PARTS LIST CAN BE IN  LOWER RIGHT
CORNER, ALLOW YOU TO FILL FROM
BOTTOM TO  TOP.

(BLOCK INSERTED)

DO NOT BLOCK H.L

| NO. | PART | REQ | MATL |
|---|---|---|---|

BASE POINT

BASE POINT

| NO. | PART | REQ | MATL |
|---|---|---|---|

**OPTION 4**

OPTION 4  ALLOWS YOU TO BUILD PARTS
LIST AS BIG AS YOU NEED BY INSERTING
OR ARRAYING AS MANY PARTS AS YOU NEED.
TO GET HEADER, EXPLODE MOST BOTTOM
ATTRIBUTE OR ADD HEADER WITH DDATTE.

DO NOT BLOCK H.L.

| | | | |
|---|---|---|---|
| | | | |
| | | | |
| | | | |
| | | | |
| | | | |
| | | | |
| − | − | − | − |
| NO. | PART | REQ | MATL |

## USING AUTOCAD

## PARTS LIST

| MARK SIGL | SCALE: FULL |
|---|---|
| DATE: 6/1/91 | FIG. #: PRB25 |

PROBLEM 26   BORDER "C" SIZE

THE SIZE OF A "C" BORDER IS 17 x 22.  LIMITS NEED TO BE CHANGED WITH THE BORDER AND TITLE BOX DESIGN.  A NEW ATTRIBUTE "TITLE" IS
USED TO FIT THE STYLE OF THE TITLE BOX.  THE SAME NAME IS USED SO YOU CAN REMEMBER THE NAME OF THE BLOCK EASIER BECAUSE THEY
PROVIDE TITLE INFORMATION.   BLOCK CONTAINS ATTRIBUTE INFORMATION: YOUR CLIENT'S NAME, YOUR NAME, TITLE OF PROBLEM,
SCALE OF DRAWING, AND DATE YOU CREATED DRAWING.   THIS TEXT INFORMATION IS KEPT ON A LAYER CALLED "BORTEXT."  THIS ALLOWS
YOU TO FREEZE THE LAYER FOR ROUGH DRAFTS AND TO SPEED UP REGENS BY REMOVING TEXT ON THE BORDER DRAWING.
"BORTEXT" LAYER IS RED TO ENSURE THAT THE SAME PEN THICKNESS IS CONSISTENT WITH THE OTHER TEXT ON THE DRAWING.
OF COURSE, WE CAN CHANGE THE COLOR ANY TIME BY CHANGING THAT LAYER'S COLOR.
IF YOU NEED TO RECALL HOW TO DO ATTRIBUTES, REFER TO PROBLEM 23 OR YOUR REFERENCE TEXT.
USED TEXT COMMAND TO PUT IN HEADERS "DATE:," "SCALE:," AND "FIG. #:."
USE STYLE COMMAND TO CREATE "I1" WITH AN ITALIC FONT TO BE USED FOR YOUR NAME IN TITLE BOX.
NOTE: SHORTCUT METHOD IS USE YOUR B SIZE BORDER.   COPY "B" BORDER AND MODIFY TO OBTAIN "C" SIZE BORDER.

ATTDEF
ATTRIBUTE MODES:

| ATTRIBUTE TAG: | CLIENT | TITLE | NAME | SCALE | DATE | DWG |
|---|---|---|---|---|---|---|
| ATTRIBUTE PROMPT: | ENTER CLIENT'S NAME | ENTER DRAWING TITLE | ENTER YOUR NAME | ENTER DRAWING SCALE | ENTER TODAY'S DATE | ENTER DRAWING # |
| DEFAULT VALUE: | INSTRUCTOR | BORDER "C" SIZE | "YOUR NAME" | FULL | 1/1/91 | 1-1 |
| START POINT: | USE "MID" & SELECT MID | USE "CENTER" & SELECT CENTER | CHANGE STYLE TO *ITALIC* PICK L.L.C. | PICK L.L.C. | PICK L.L.C. | PICK L.L.C. |
| HEIGHT <.125>: | .20 | .20 | .125 | <CR> | <CR> | <CR> |
| ROTATION ANGLE <0>: | <CR> | <CR> | <CR> | <CR> | <CR> | <CR> |

BLOCK ATTRIBUTES
BLOCK NAME: TITLE     THIS RECALLS THE ATTRIBUTE FOR YOUR TITLE BOX INFORMATION
INSERTION POINT: 0,0    SAME AS BORDER ORIGIN
SELECT:   PICK IN ORDER (CLIENT, TITLE, NAME, SCALE, DATE, DWG #)
          THIS CAUSES THE ATTRIBUTES TO ASK FOR INFORMATION IN THAT ORDER.

AFTER ALL UPDATES HAVE BEEN PERFORMED, SAVE THE NEW BORDER AS "BORC."

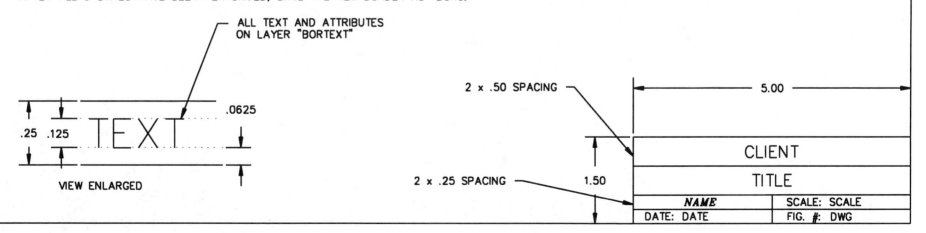

PROBLEM 26   BORDER "C" SIZE

| USING AUTOCAD | |
|---|---|
| BORDER "C" SIZE | |
| MARK SICL | SCALE: FULL |
| DATE: 6/1/91 | FIG. #:  PRB 26 |

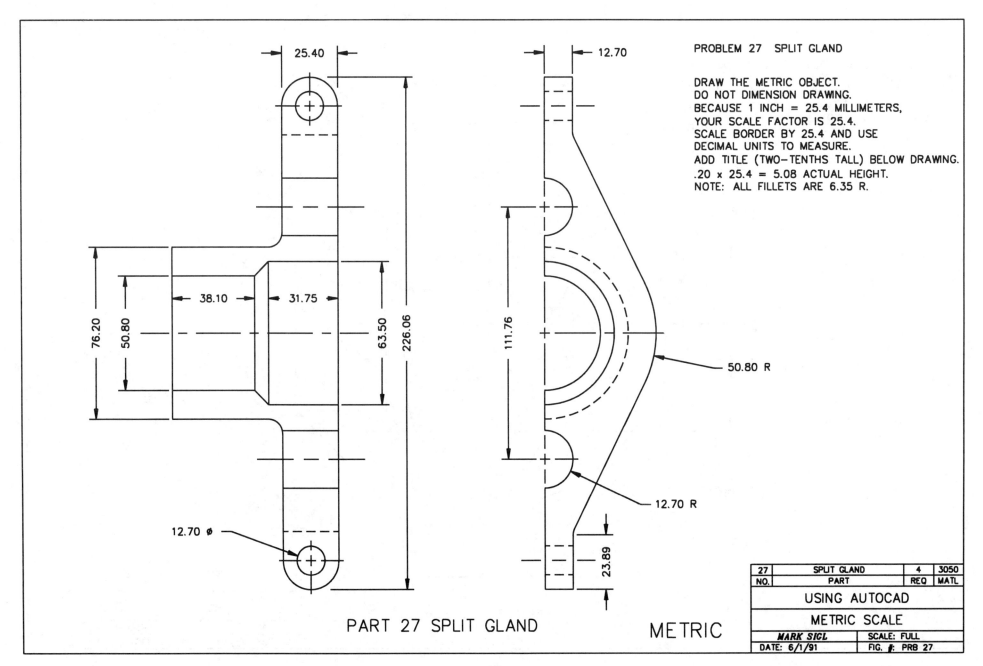

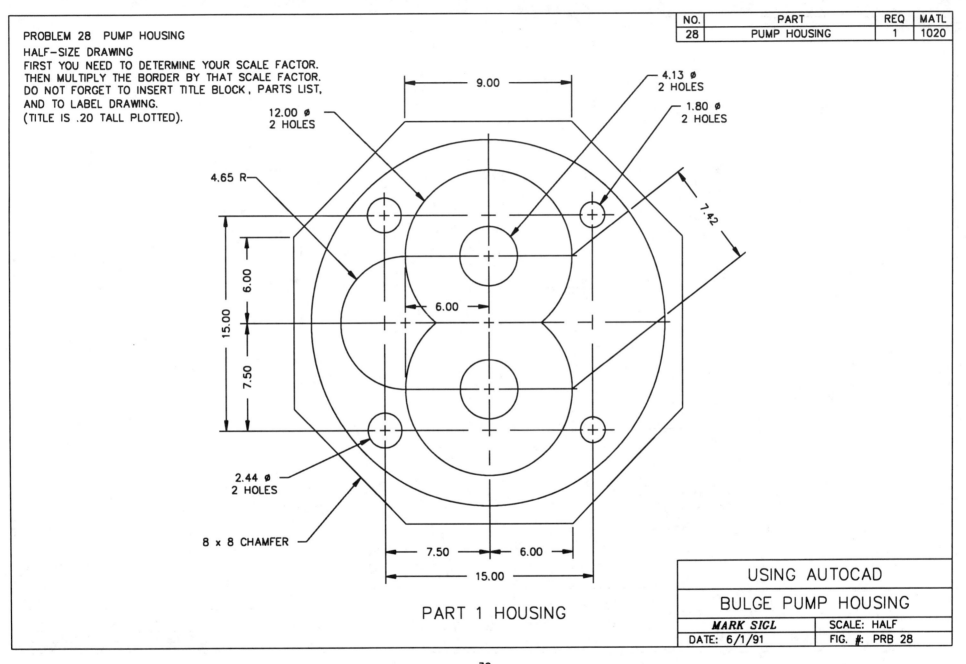

PROBLEM 29
HING PLATE

DRAW OBJECT AND DO NOT DIMENSION. REMEMBER
ABOUT YOUR SCALE FACTOR.
IF YOU NEED HELP FINDING LOCATION OF LINES, PROJECTION
THEM BETWEEN VIEWS (USE 45° LINE FOR TOP AND RIGHT
SIDE VIEW PROJECTION).

| NO. | PART | REQ | MATL |
|---|---|---|---|
| 29 | HING PLATE | 6 | 1020 |

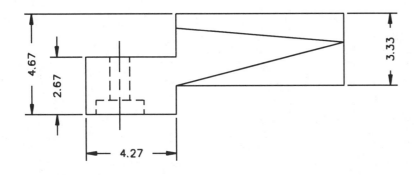

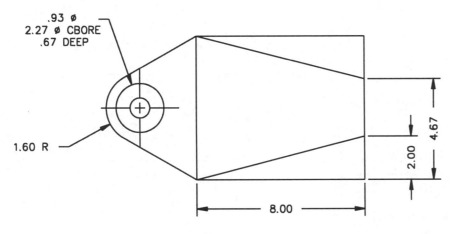

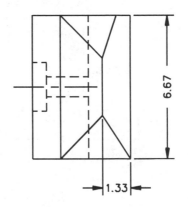

PART 29 HING PLATE

| USING AUTOCAD ||
|---|---|
| HING PLATE ||
| *MARK SIGL* | SCALE: 3/8" = 1" |
| DATE: 6/1/91 | FIG. #: PRB 29 |

37

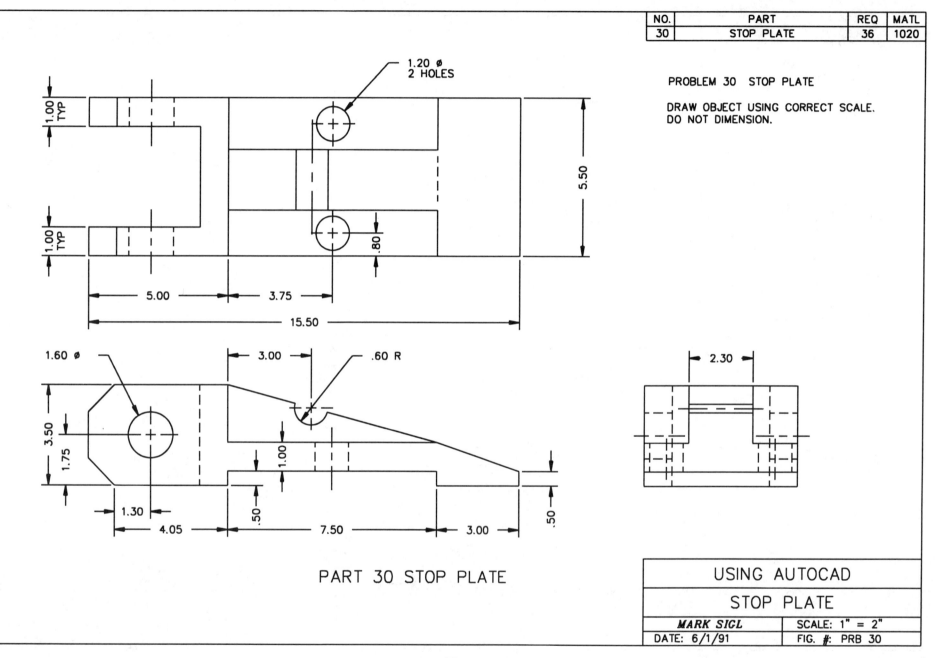

# CHAPTER 4
# DIMENSIONING

THIS CHAPTER COVERS THE USE OF DIMENSIONING COMMANDS AND VARIABLES.  DIMENSIONING PROVIDES YOU WITH A SET OF ACCURATE MEASUREMENTS.  THIS ALLOWS YOU TO PERFORM ACCURATE COST ESTIMATES, PROVIDE FOR ACCURATE MANUFACTURED PARTS, AND PROVIDE INSTRUCTIONS FOR BUILDING AND LAYOUT MEASUREMENTS.  PROPER DIMENSIONING IS IMPORTANT AS ACCURATE DRAWING.

PROBLEM                                                              TOPIC

31      EXERCISE EXPOSES YOU TO "DIM VARIABLES."  THE MORE FAMILIAR YOU ARE WITH THE DIMENSIONING
        VARIABLES, THE FASTER YOU CAN DIMENSION A DRAWING.

32      USE TWO OF THE MOST COMMONLY USED DIMENSIONING COMMANDS — VERTICAL AND HORIZONTAL DIMENSIONING.  THESE ARE THE
        MOST BASIC AND EASIEST TO USE.

33      THIS EXERCISE EXPLORES WAYS TO DIMENSION ARCS AND CIRCLES USING LEADER, DIAMETER, AND RADIUS DIMENSIONING COMMANDS.

34      MANIPULATING THE AUTO DIMENSIONING MODE AND UTILIZING BASELINE DIMENSIONING IS THE BASIS OF THIS EXERCISE.
        YOU NEED TO PLAN HOW YOU ARE GOING TO LAY OUT THE DIMENSIONS SO YOU DO NOT CROSS OVER OTHER DIMENSIONS.

35      EXERCISE EXPOSES YOU TO CONTINUOUS DIMENSIONING AND HOW TO USE THE AUTO DIMENSIONING FEATURE.  REMEMBER
        YOU WILL NEED TO CHANGE A FEW DIMENSIONING VARIABLES EVERY NOW AND THEN.

36      EXERCISE DEMONSTRATES THE DIFFERENCE BETWEEN ALIGN AND ROTATE DIMENSIONING.  YOU DECIDE WHICH ONE
        TO USE TO MEET YOUR NEEDS.

37      IN UNDERSTANDING HOW TO DO LIMIT DIMENSIONING, THE HARD PART IS DETERMINING THE UPPER AND LOWER LIMITS UNLESS
        YOU HAVE AN ENGINEER TO ESTABLISH THE TOLERANCES NEEDED.

38      EXERCISE COVERS TOLERANCE DIMENSIONING AND EXPOSES YOU TO CHANGING THE UPPER AND LOWER RANGES AS
        SPECIFIED BY THE ENGINEER.

39      EXERCISE DEMONSTRATES THE EASE OF USING THE METRIC SYSTEM VERSUS THE ENGLISH DIMENSIONING SYSTEM.

40      EXERCISE COVERS FRACTIONAL DIMENSIONING.

## MATCHING DIM VARIABLES

PROBLEM 31
DIM VARIABLES

A. DIMTSZ
B. DIMASZ
C. DIMEXE
D. DIMEXO
E. DIMTAD
F. DIMSE1 (ON)
G. DIMSE2 (ON)
H. DIMSE1 (OFF)
I. DIMSE2 (OFF)

J. DIMTOD
K. DIMTXT
L. UNITS
M. DIMTOH
N. DIMTIH
O. DIMTIX
P. DIMDLE
Q. DIMDLI
R. DIMTXE

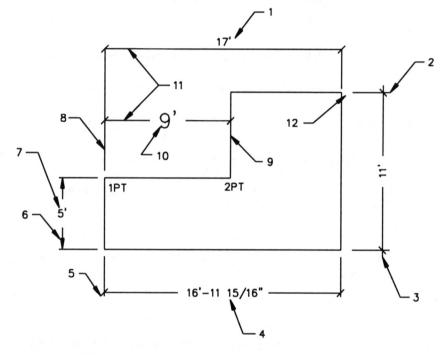

MATCH THE NUMBER TO THE CORRECT DIM VARIABLE.

1. _____
2. _____
3. _____
4. _____
5. _____
6. _____

7. _____
8. _____
9. _____
10. _____
11. _____
12. _____

| USING AUTOCAD ||
|---|---|
| DIM VARIABLES ||
| *MARK SIGL* | SCALE: FULL |
| DATE: 6/1/91 | FIG. #: PRB 31 |

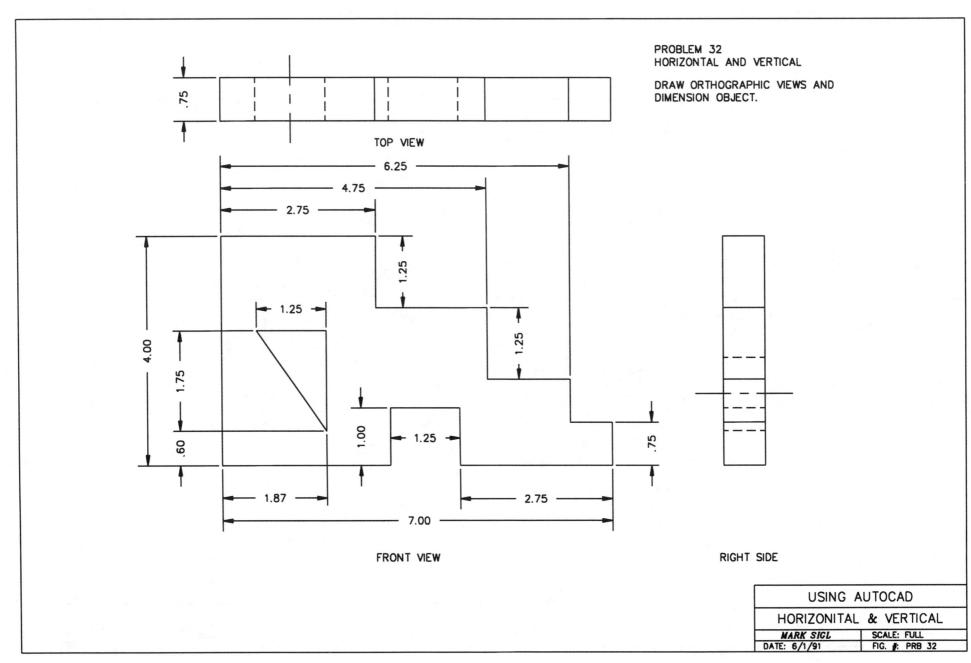

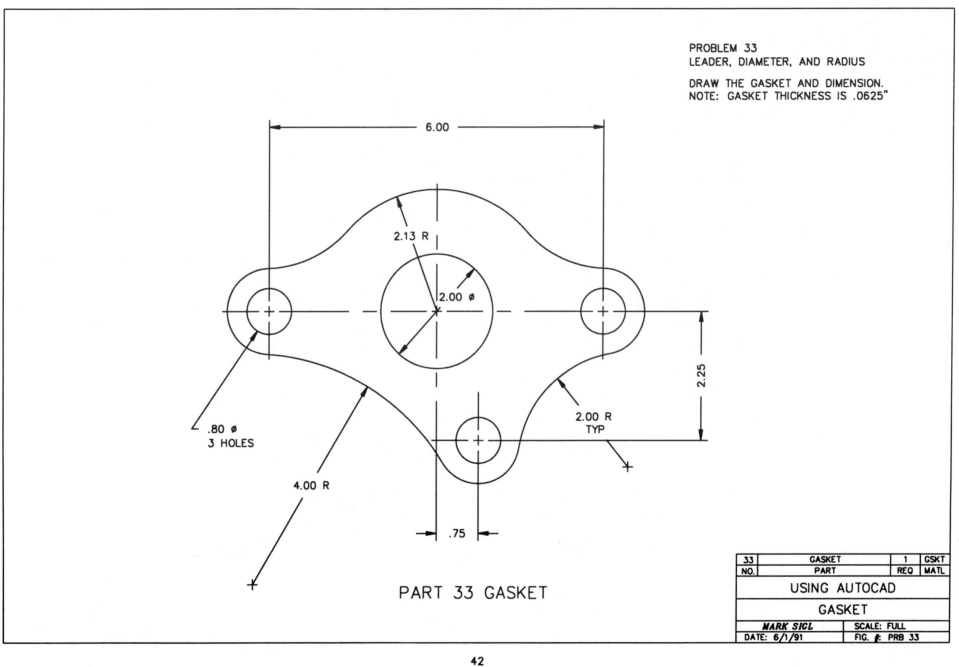

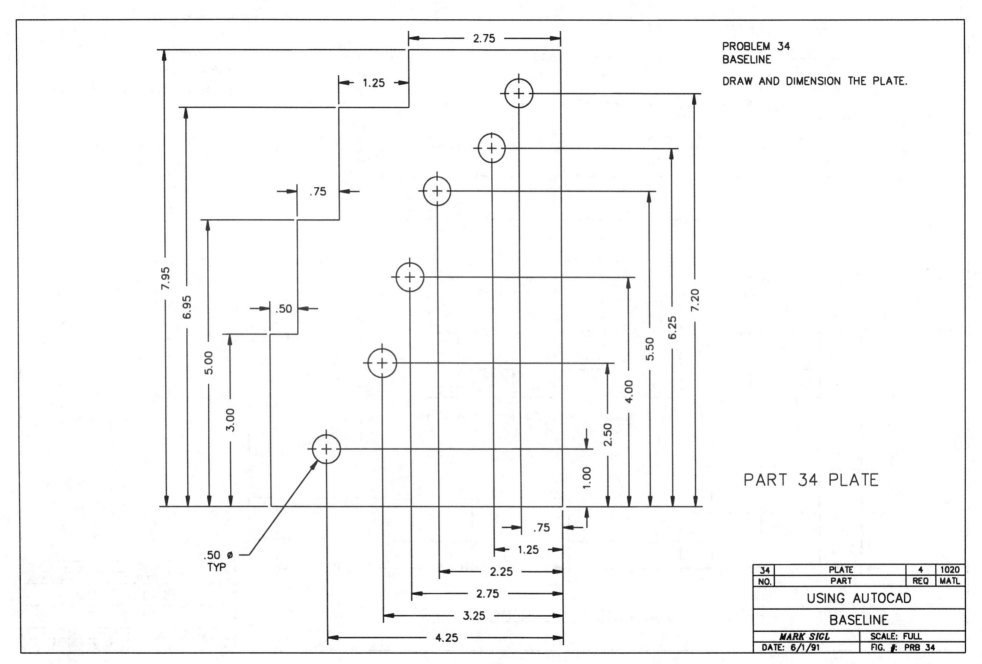

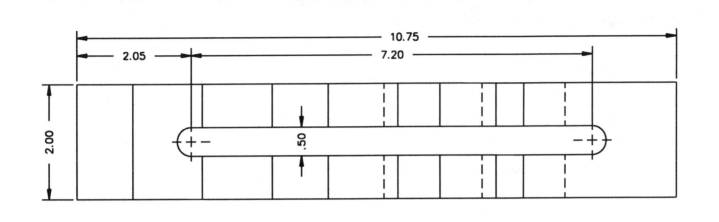

PROBLEM 35
CONTINUOUS

DRAW AND DIMENSION THE BLOCK.

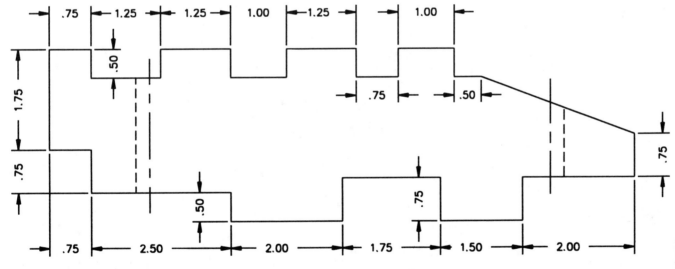

PART 35 SLIDE BLOCK

| 35 | BLOCK | 4 | 1020 |
|---|---|---|---|
| NO. | PART | REQ | MATL |

USING AUTOCAD

CONTINUOUS

| MARK SIGL | SCALE: FULL |
|---|---|
| DATE: 6/1/91 | FIG. #: PRB 35 |

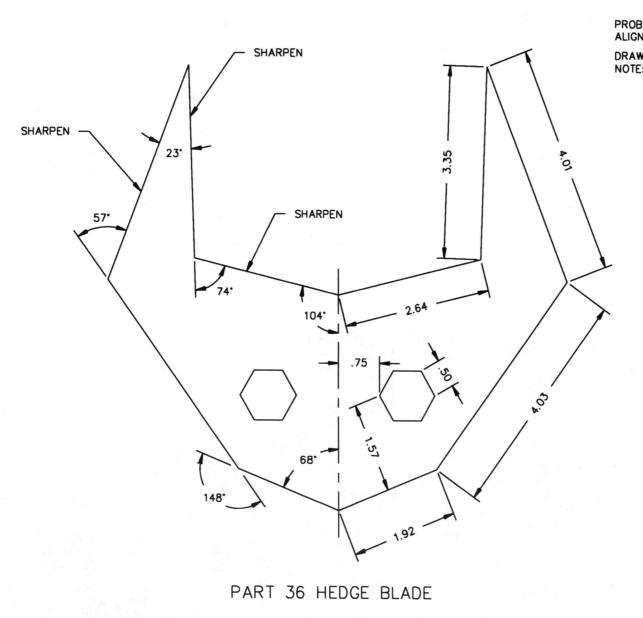

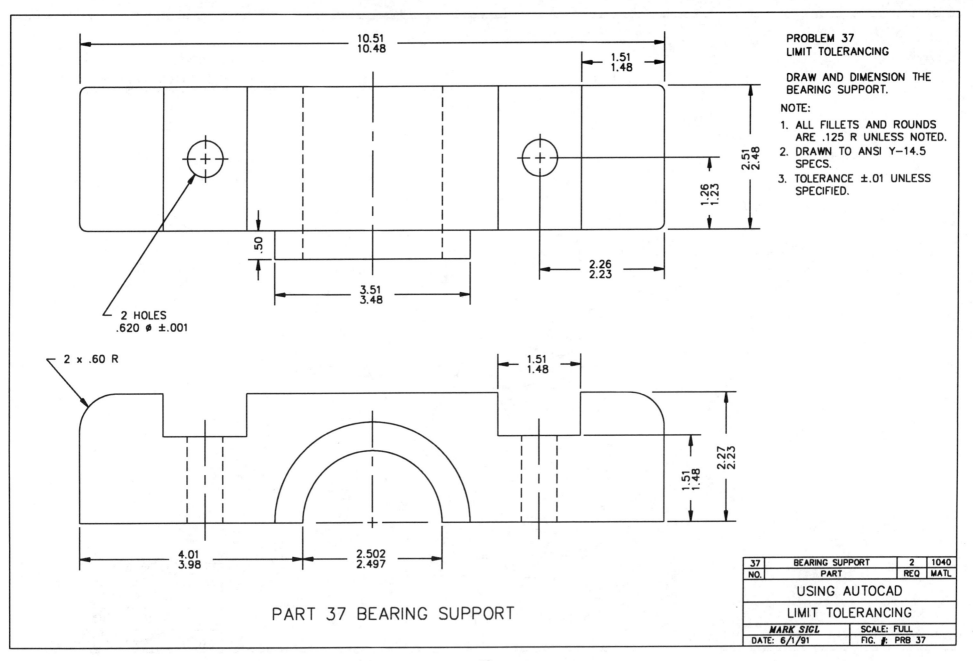

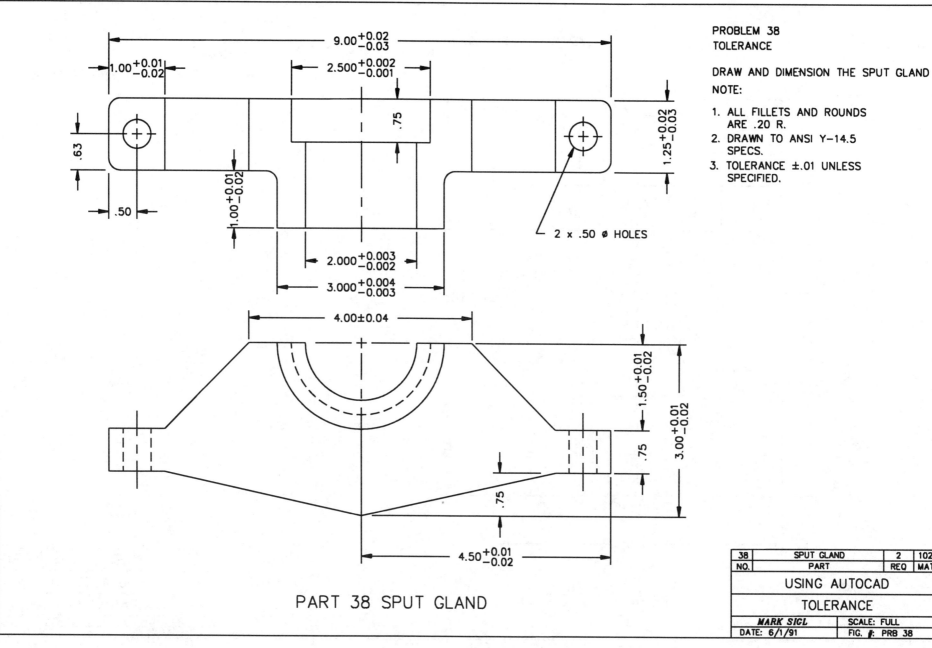

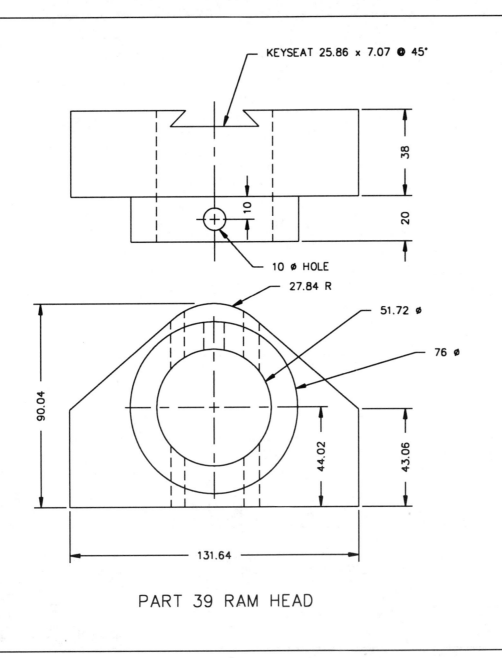

PROBLEM 39
RAM HEAD

DRAW THE RAM HEAD AND DIMENSION.

PART 39 RAM HEAD

METRIC

| 39 | RAM HEAD | 1 | 1020 |
|---|---|---|---|
| NO. | PART | REQ | MATL |

USING AUTOCAD

RAM HEAD

| *MARK SIGL* | SCALE: FULL |
|---|---|
| DATE: 6/1/91 | FIG. #: PRB 39 |

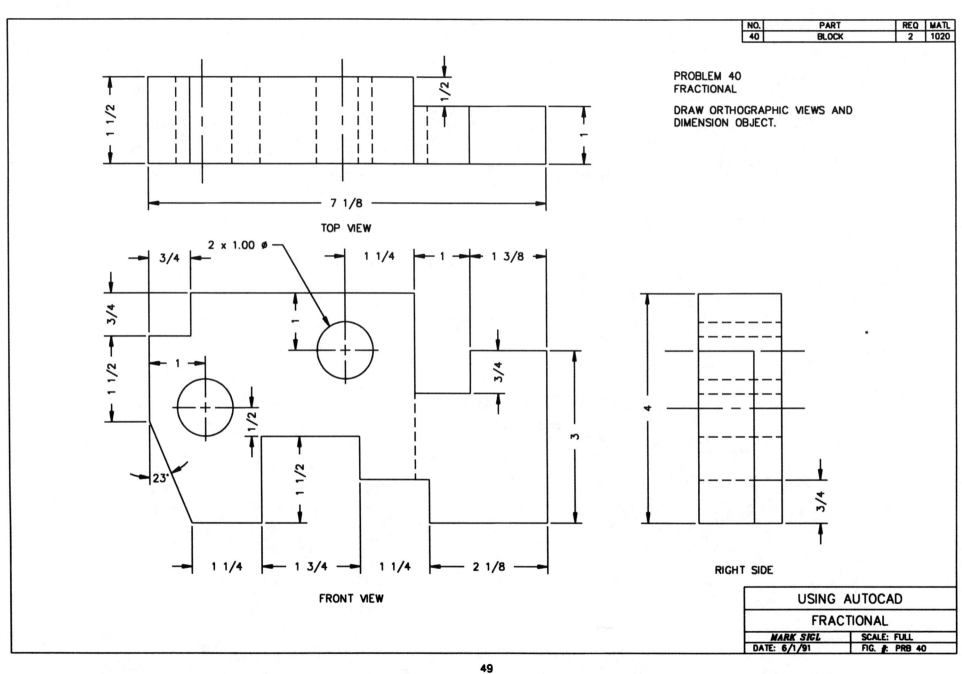

# CHAPTER 5
# ORTHOGRAPHIC

THIS CHAPTER COVERS MULTIVIEW DRAWINGS OFTEN CALLED "ORTHOGRAPHIC PROJECTION." BY LOOKING AT A DRAWING FROM THE FRONT VIEW, TOP VIEW, AND RIGHT SIDE VIEW, ONE CAN GET A CLEAR PICTURE OF HOW THE OBJECT WILL LOOK. IN THIS SYSTEM, THE VIEWS OF AN OBJECT ARE PROJECTED PERPENDICULARLY ONTO PROJECTION PLANES WITH PARALLEL PROJECTORS. IN OTHER WORDS, THERE IS A TRANSPARENT SCREEN BETWEEN YOU AND OBJECT. OBJECT IS PROJECTED FORWARD ONTO THE SCREEN. IN THIS CHAPTER YOU DETERMINE HOW MANY VIEWS ARE NECCESSARY TO DRAW, OBTAIN THE THIRD VIEW FROM THE OTHER TWO, DRAW THE THREE VIEWS FROM AN ISOMETRIC DRAWING, AND DETERMINE WHAT THE OBJECT LOOKS LIKE FROM AUXILIARY AND PARTIAL VIEWS. ALSO, YOU WILL GAIN MORE EXPERIENCE IN DIMENSIONING.

| PROBLEM | TOPIC |
|---|---|
| 41 | WITH THE BASE BEARING DRAWING YOU DRAW AN ORTHOGRAPHIC TWO-VIEW DRAWING WITH DIMENSIONING. |
| 42 | THE BEARING BRACKET IS A TWO-VIEW DRAWING WITH FILLETS, TANGENT LINES, AND DIMENSIONING. |
| 43 | ON THE SLIDE BRACKET YOU ARE EXPOSED TO ROUND OUTS AND ARCS TANGENT TO OTHER ARCS. |
| 44 | WITH THE SUPPORT BRACKET YOU DRAW THE THIRD VIEW OBTAINED FROM A TWO-VIEW DRAWING. BE CAREFUL WITH TANGENCY OF ARC. |
| 45 | GIVEN AN ISOMETRIC DRAWING OF A T-GUIDE, DRAW THE NECESSARY ORTHOGRAPHIC VIEWS AND DIMENSION. |
| 46 | ISOMETRIC DRAWING OF THE UNIVERSAL JOINT OFFERS A CHALLENGE IN VISUALIZATION AND IN OBTAINING THE ORTHOGRAPHIC VIEWS. |
| 47 | BEARING SADDLE GIVES YOU MORE EXPERIENCE IN PREPARING ORTHOGRAPHIC VIEWS. |
| 48 49 50 | LAST THREE DRAWINGS ARE DESIGNED TO GIVE FURTHER EXPERIENCE WITH CREATING ORTHOGRAPHIC PROJECTIONS AND ALLOWS YOU TO BECOME MORE PROFICIENT WITH MANIPULATING AND PLACING DIMENSIONS. |

PROBLEM 41
BASE BRACKET

DRAW ORTHOGRAPHIC VIEWS
AND DIMENSION.

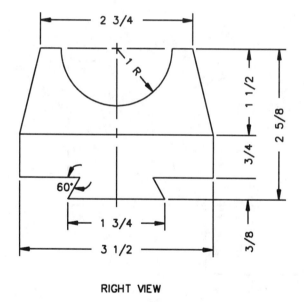

| 21 | BRACKET | 2 | 1360 |
|---|---|---|---|
| NO. | PART | REQ | MATL |

USING AUTOCAD

BASE BRACKET

| MARK SIGL | SCALE: FULL |
|---|---|
| DATE: 6/1/91 | FIG. #: PRB 41 |

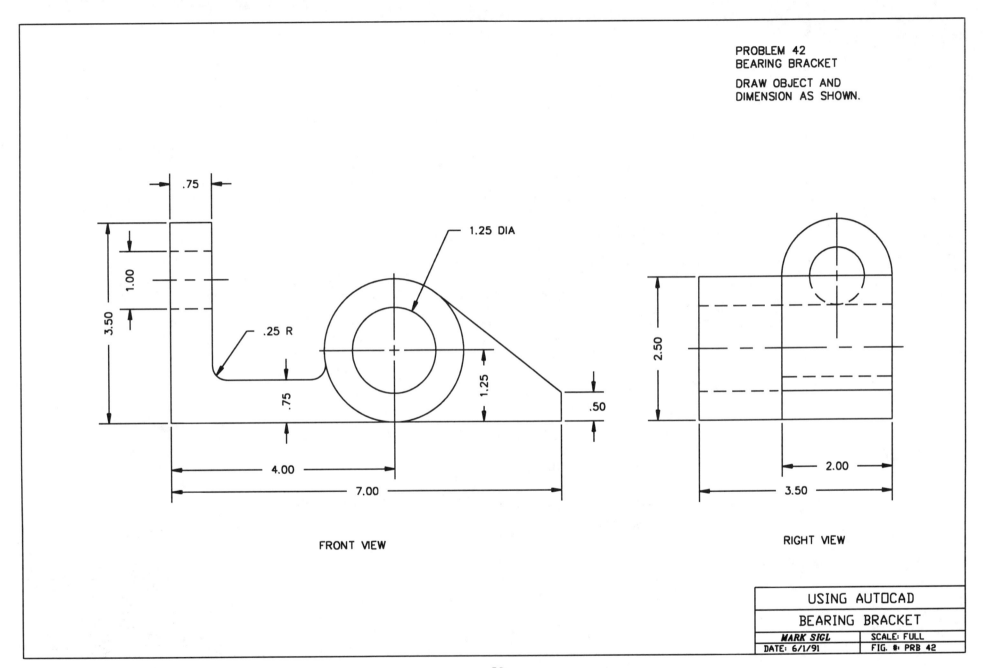

PROBLEM 43
SLIDE BRACKET

DRAW ORTHOGRAPHIC VIEWS
AND DIMENSION.
NOTE: ALL FILLETS ARE .125 RADIUS.

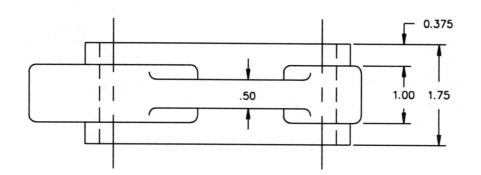

TOP VIEW

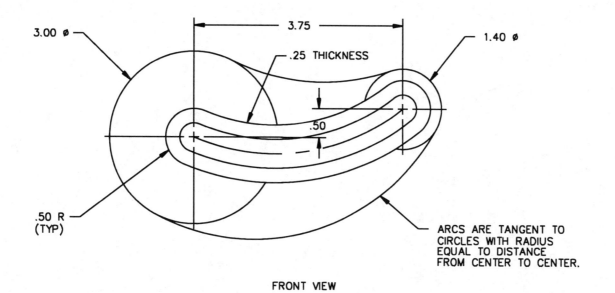

FRONT VIEW

| 43 | SLIDE BRACKET | 4 | 1020 |
|---|---|---|---|
| NO. | PART | REQ | MATL |

USING AUTOCAD
SLIDE BRACKET

| MARK SIGL | SCALE: FULL |
|---|---|
| DATE: 6/1/91 | FIG. #: PRB 43 |

PROBLEM 44
SUPPORT BRACKET

DRAW THREE VIEWS AND
DIMENSION.

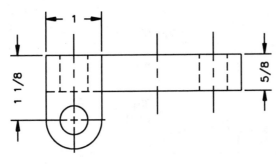

TOP VIEW

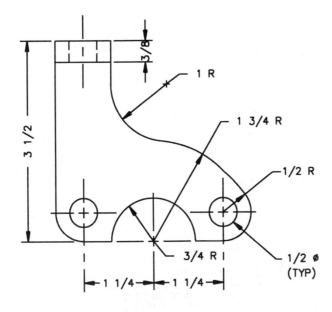

FRONT VIEW

RIGHT VIEW

| USING AUTOCAD |||
|---|---|---|
| SUPPORT BRACKET |||
| *MARK SIGL* | SCALE: FULL ||
| DATE: 6/1/91 | FIG. #: PRB 44 ||

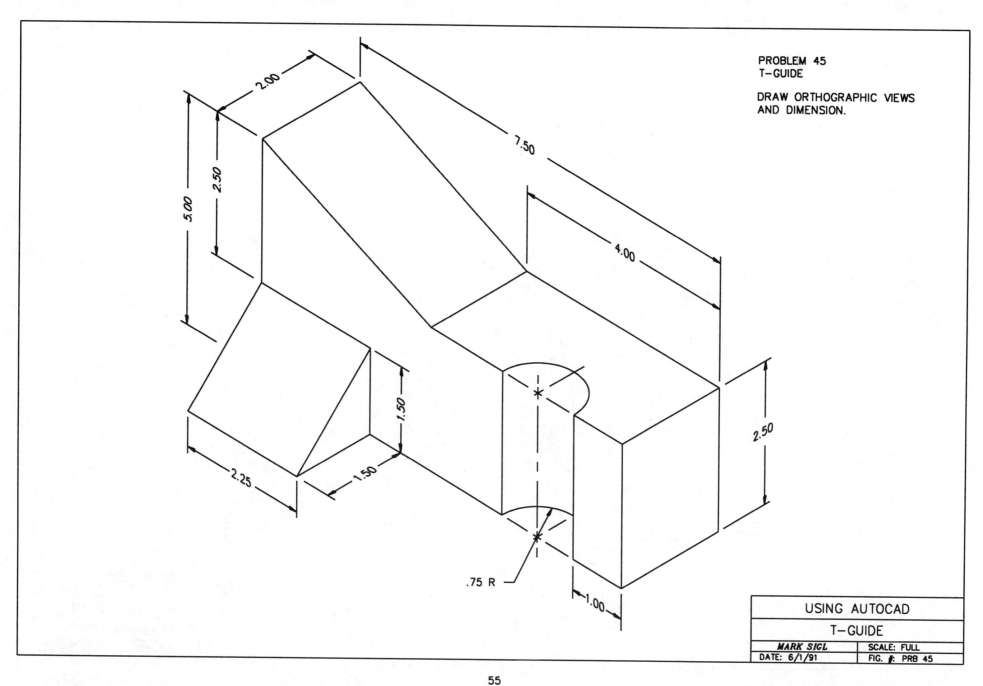

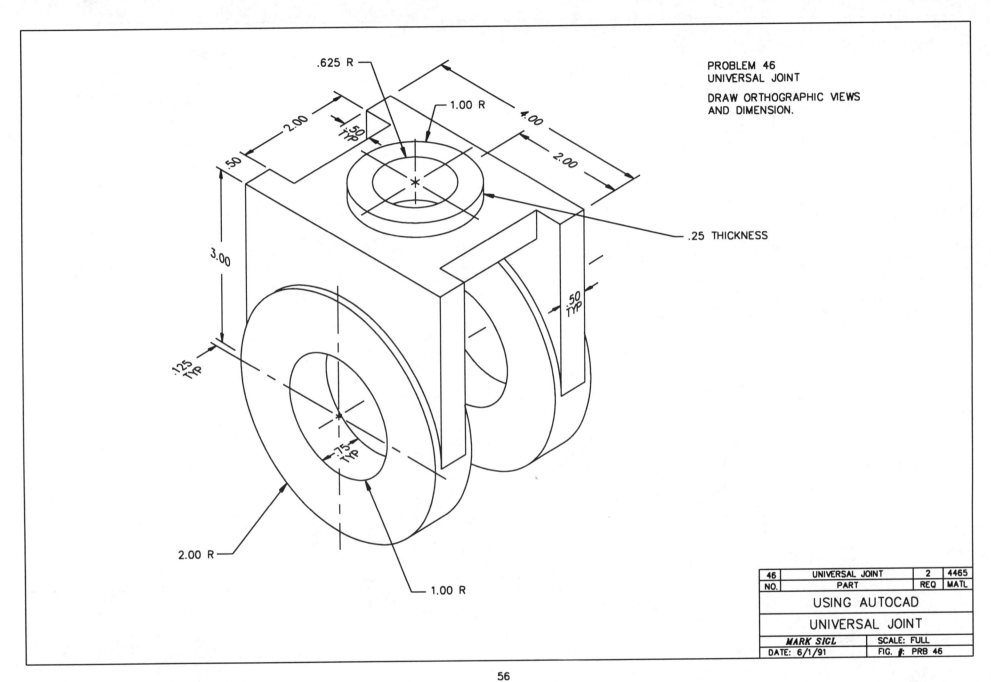

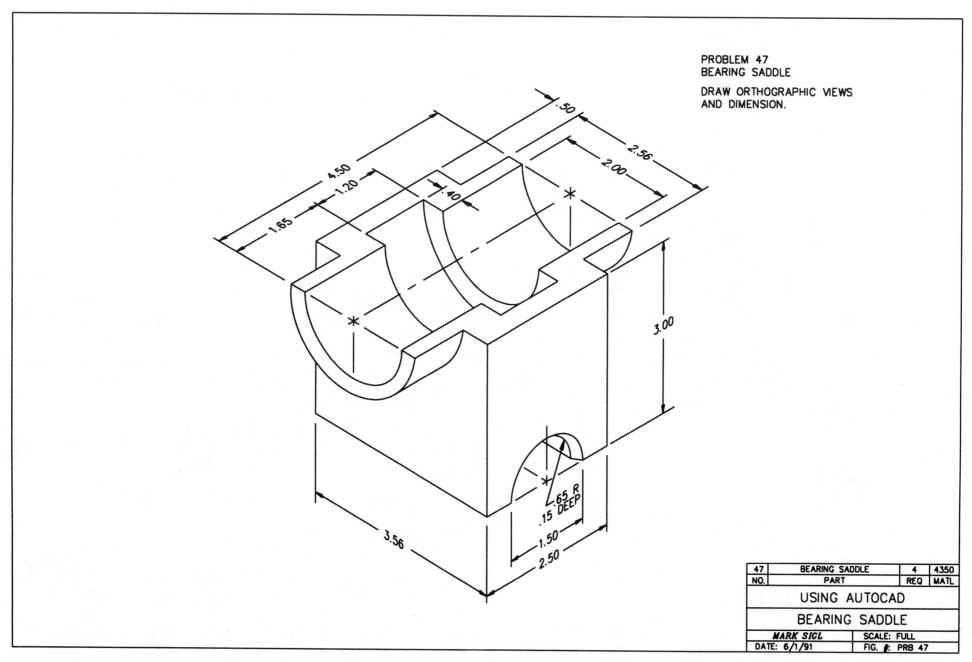

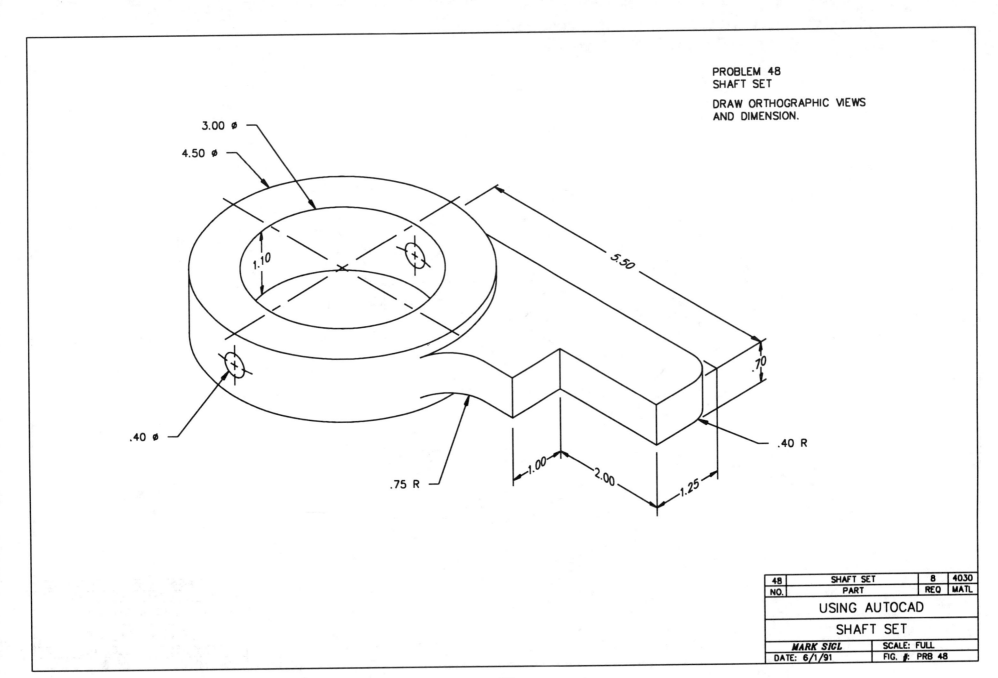

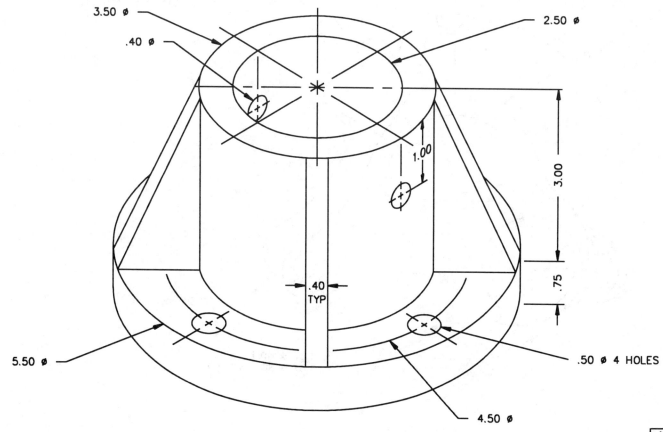

| NO. | PART | REQ | MATL |
|-----|------|-----|------|
| 50 | WING POST | 12 | 1020 |

PROBLEM 50
WING POST

GIVEN AN OBLIQUE VIEW, DRAW NECESSARY
ORTHOGRAPHIC VIEWS AND DIMENSION.
NOTE: OVERALL DEPTH IS 7.33".

2.67 ⌀

6.66 ⌀

8.43 ⌀

8.00 R

.93 x .93 CHAMFER

2.00 DEPTH

1.07 ⌀ HOLE
1.87 ⌀ SF .33 DEEP

4.80 ⌀

4.00 ⌀
7.33 THROUGH

8.00 R

PART 50 WING POST

| USING AUTOCAD | |
|---|---|
| WING POST | |
| *MARK SIGL* | SCALE: 3/8" = 1" |
| DATE: 6/1/91 | FIG. #: PRB 50 |

# CHAPTER 6
# ISOMETRIC & OBLIQUE

THIS CHAPETR COVERS CREATING ISOMETRIC AND OBLIQUE DRAWINGS.  THESE PICTORIAL DRAWINGS ARE USED TO EXPRESS HOW THE OBJECT WOULD LOOK THREE-DIMENSIONALLY.  NOTICE THAT THE HIDDEN LINES ARE REMOVED FOR CLARITY.  OBJECT IS GIVEN DEPTH BY REVOLVING OR PROJECTING THE OBJECT LINES NOT PERPENDICULAR (NOT 90°) TO THE PROJECTION PLANE.  THIS FORM OF ILLUSTRATION EXPOSES THREE SIDES OF THE OBJECT AT ANY GIVEN TIME.  BECAUSE THIS IS A BEGINNING PROBLEM, WE WILL DRAW THE OBJECTS AND NOT DIMENSION THEM, BECAUSE MOST CAD PROGRAMS REQUIRE YOU TO CUSTOMIZE ARROWHEADS AND TEXT FOR ISOMETRIC AND OBLIQUE DIMENSIONING.

| PROBLEM | TOPIC |
|---|---|
| 51 | T-GUIDE IS A SIMPLE ISOMETRIC THAT INCLUDES AN ELLIPSE AND INCLINED SURFACES TO FIND. |
| 52 | WITH THE STOP BRACKET YOU ARE EXPOSED TO INCLINED SURFACES AND HOLES THROUGH AN OBJECT. |
| 53 | THE FIXTURE END BLOCK HAS INCLINED SURFACES AND CORNERS TO LOCATE, GIVING YOU MORE EXPERIENCE WITH ISOMETRIC DRAWINGS AND PLANES. |
| 54 | WITH THE HOLD-DOWN CLAMP YOU ARE GIVEN AN ORTHOGRAPHIC VIEW AND YOU NEED TO PRODUCE ISOMETRIC AND OBLIQUE VIEWS. |
| 55 | WITH THE SLIDE BLOCK IT WILL KEEP YOUR INTEREST WITH LEDGES, ELLIPSES, AND LOCATIONS OF HOLES. |
| 56 | LUG DRAWING GIVES YOU EXPOSURE TO DRAWING AN ISOMETRIC SECTION AND AN OBLIQUE SECTION VIEW. |
| 57 | SHAFT SET EXPOSES YOU TO ISOMETRIC ROUNDS AND FILLETS AND A RECESSED SURFACE. |
| 58 | WITH THE LIFT BLOCK YOU ADD ELLIPSES TO GRAPHICALLY ILLUSTRATE THE ROUNDS.  HINT: BLOCK ELLIPSE AND MEASURE THE LINE TO PLACE THE ARCS.  (GOOD LUCK) |
| 59 | UNIVERSAL JOINT IS ULTIMATE ISOMETRIC PROBLEM DEALING WITH ELLIPSES, RECESSED SURFACES, AND THROUGH HOLES. |
| 60 | BEARING BLOCK OFFERS OFFSET ELLIPSES THAT WILL CHALLENGE YOU. |

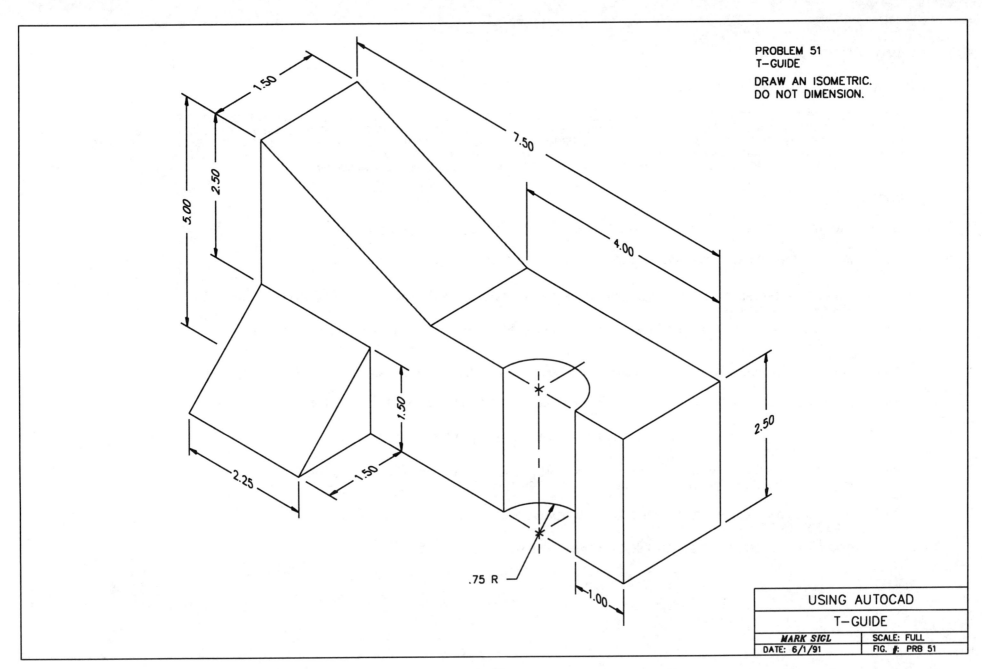

PROBLEM 51
T-GUIDE

DRAW AN ISOMETRIC.
DO NOT DIMENSION.

| USING AUTOCAD ||
| T-GUIDE ||
| *MARK SIGL* | SCALE: FULL |
| DATE: 6/1/91 | FIG. #: PRB 51 |

PROBLEM 52
STOP BRACKET

DRAW AN ISOMETRIC.
DO NOT DIMENSION.

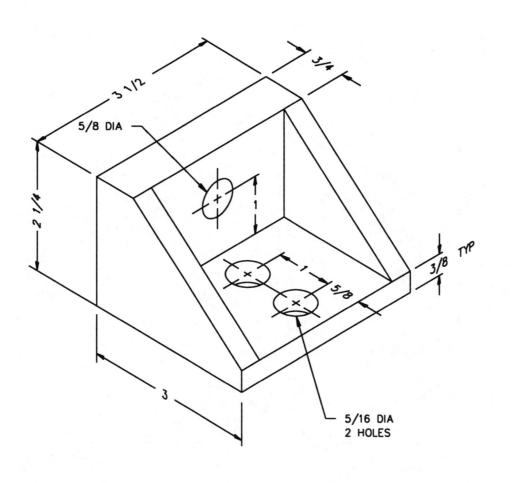

| 52 | STOP BRACKET | 2 | 2506 |
|---|---|---|---|
| NO. | PART | REQ | MATL |

USING AUTOCAD

STOP BRACKET

| *MARK SIGL* | SCALE: FULL |
|---|---|
| DATE: 6/1/91 | FIG. #: PRB 52 |

PROBLEM 53
FIXTURE END BLOCK

DRAW AN ISOMETRIC VIEW.
DO NOT DIMENSION.

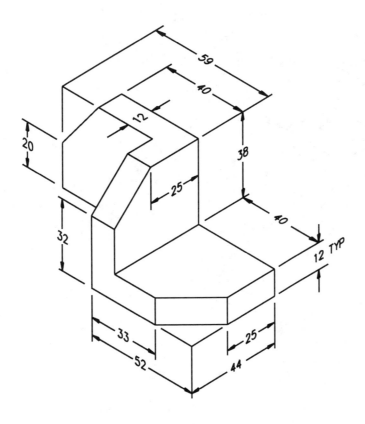

(METRIC)

| 53 | FIXTURE END | 6 | 2340 |
|---|---|---|---|
| NO. | PART | REQ | MATL |

USING AUTOCAD

FIXTURE END BLOCK

| MARK SIGL | SCALE: FULL |
|---|---|
| DATE: 6/1/91 | FIG. #: PRB 53 |

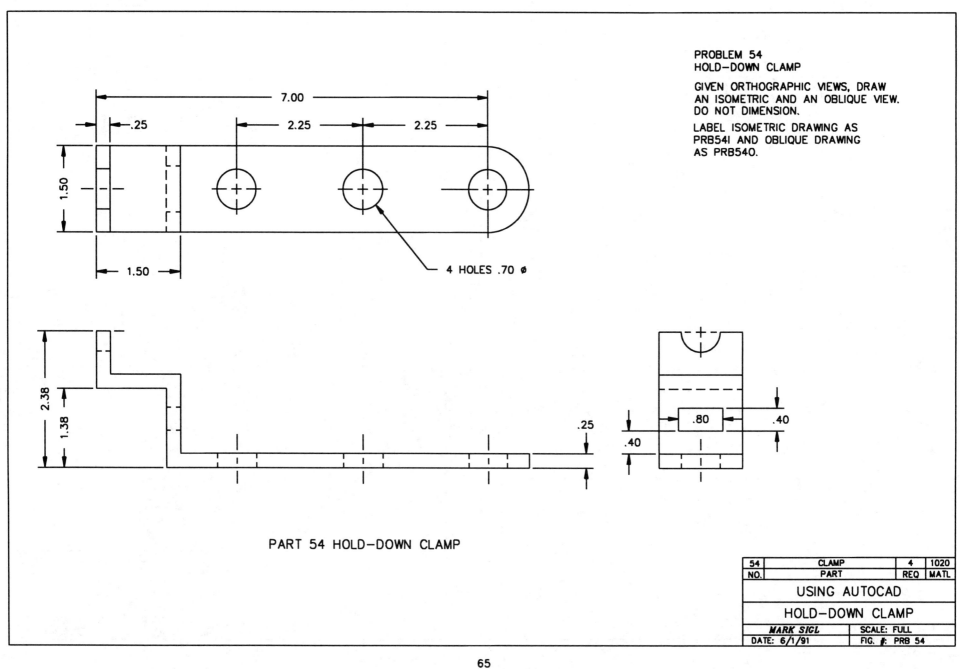

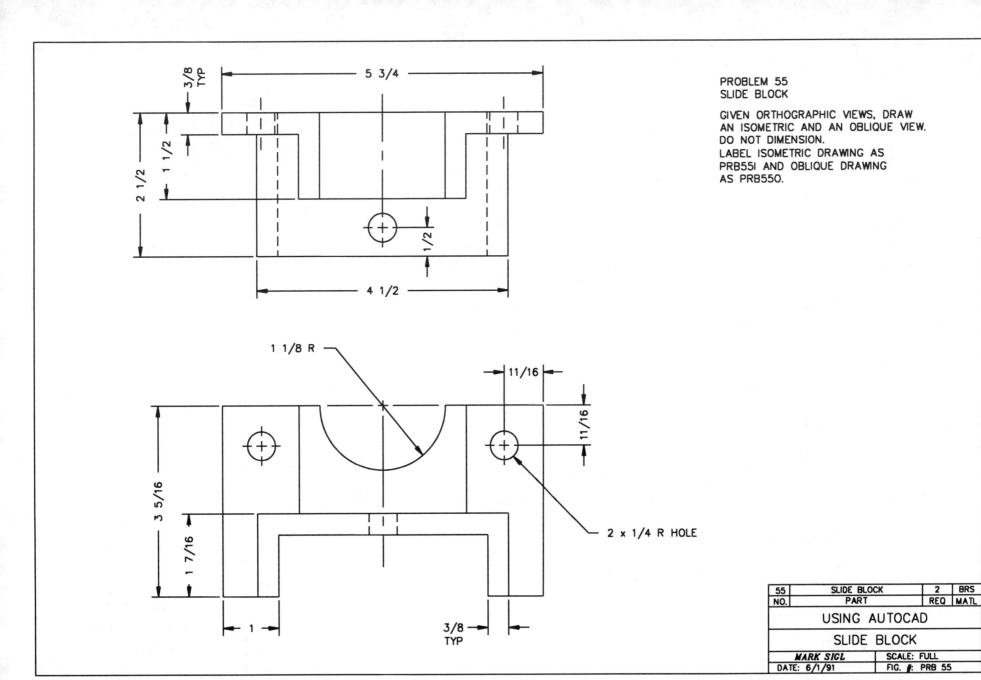

PROBLEM 56
LUG

GIVEN ORTHOGRAPHIC VIEW AND
SECTION VIEW, DRAW AN ISOMETRIC AND
AN OBLIQUE VIEW.  DO NOT DIMENSION.

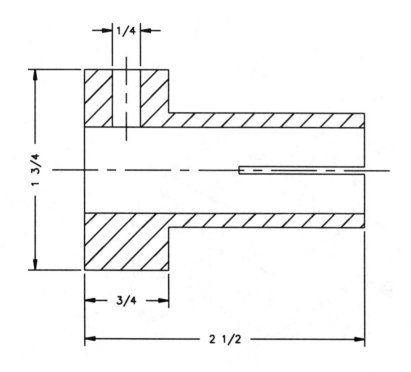

SECTION A-A

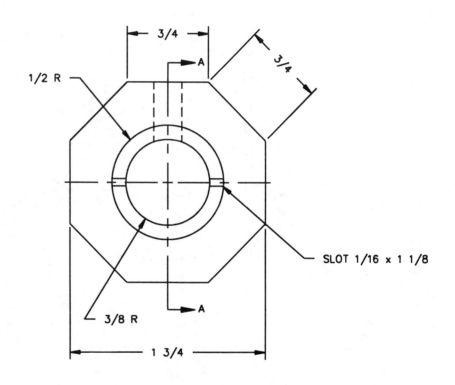

| 56 | LUG | 8 | 1800 |
|---|---|---|---|
| NO. | PART | REQ | MATL |

USING AUTOCAD

LUG

| MARK SIGL | SCALE: 2=1 |
|---|---|
| DATE: 6/1/91 | FIG. #: PRB 56 |

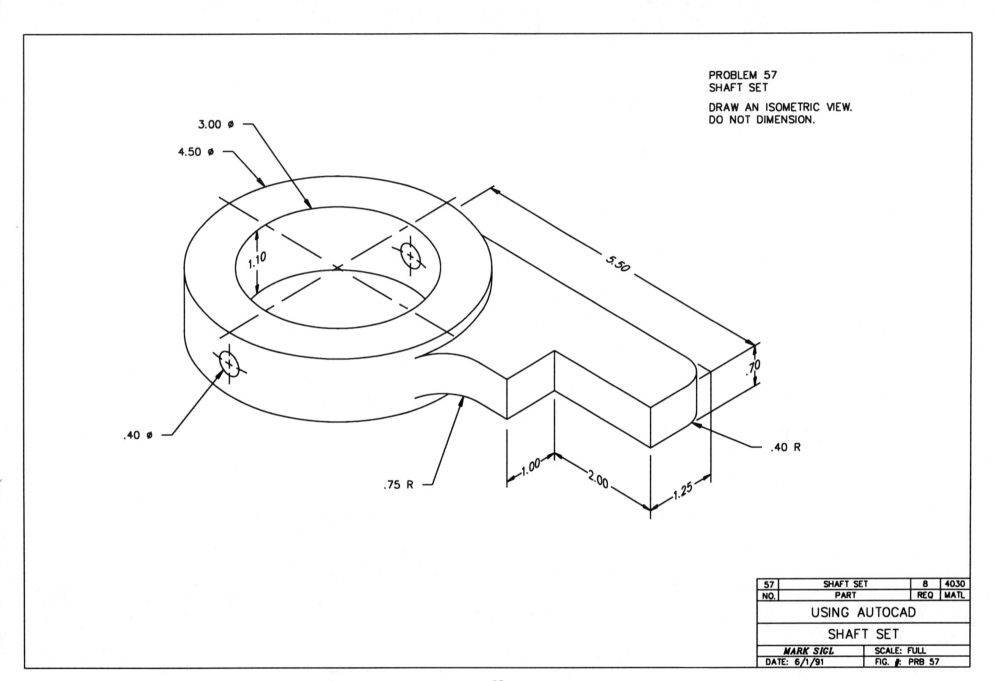

PROBLEM 58
LIFT BLOCK

DRAW AN ISOMETRIC.
DO NOT DIMENSION.
NOTE: ALL FILLETS ARE .125 R.

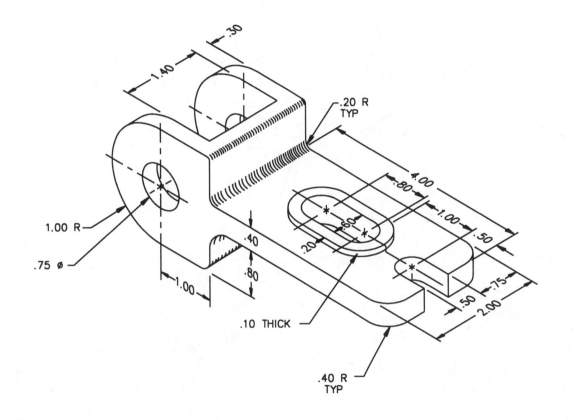

| 58 | LIFT BLOCK | 2 | 1020 |
|----|------------|---|------|
| NO. | PART | REQ | MATL |

USING AUTOCAD

LIFT BLOCK

| MARK SIGL | SCALE: FULL |
|-----------|-------------|
| DATE: 6/1/91 | FIG. #: PRB 58 |

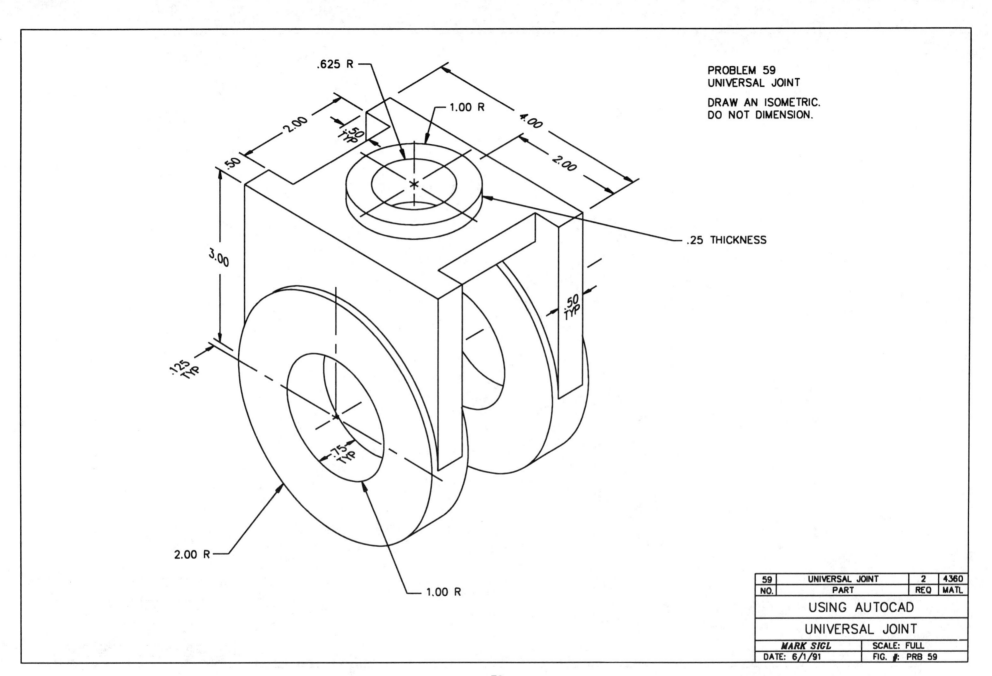

PROBLEM 60
BEARING BLOCK

DRAW AN ISOMETRIC VIEW.
FOR EXTRA CREDIT DRAW
AN OBLIQUE VIEW.
DO NOT DIMENSION.

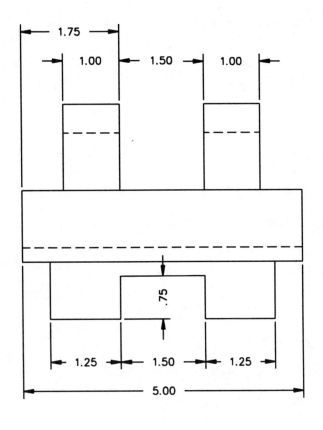
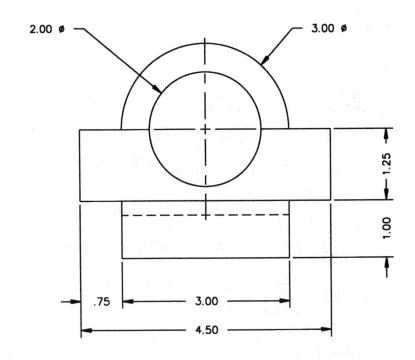

| 60 | BEARING BLOCK | 2 | 3150 |
|----|---------------|---|------|
| NO. | PART | REQ | MATL |

USING AUTOCAD
BEARING BLOCK

| MARK SIGL | SCALE: FULL |
|-----------|-------------|
| DATE: 6/1/91 | FIG. #: PRB 60 |

# CHAPTER 7
# SECTIONS & HATCHING

THIS CHAPTER COVERS SECTIONING IN WHICH HATCHING IS USED TO INDICATE THE SURFACE OF THE OBJECT THAT LIES ON THE CUTTING PLANE. SECTIONING IS USED TO LOOK AT HIDDEN COMPONENTS INSIDE AN OBJECT THAT WOULD OTHERWISE BE UNCLEAR OR CONFUSING WHEN REPRESENTED WITH HIDDEN LINES.  USE THE TRIM COMMAND TO HELP YOU DEFINE THE AREA TO HATCH.  ONCE THE AREA IS HATCHED, USE THE FILLET COMMAND TO CLOSE THE OPENINGS.  THE SECTION LINES ARE USUALLY DRAWN AT 45°, OR ANY OTHER STANDARD ANGLE SUCH AS 30° OR 60°.  THE SPACING OF SECTION LINES ARE USUALLY 1/16" TO 1/8" APART OR WIDER IF THERE IS A LOT OF AREA TO SECTION. SOME CAD PROGRAMS HAVE PREMADE HATCH PATTERNS TO REPRESENT COMMON MATERIALS USED.  THESE PATTERNS ARE USUALLY ANSI STANDARDS. MOST CAD PROGRAMS WILL LET YOU CREATE OR CUSTOMIZE THE HATCH PATTERN THAT IS AVAILABLE.

| PROBLEM | TOPIC |
|---|---|
| 61 | HUB BASE WILL BE A HALF SECTION, REMOVING A QUARTER OF THE OBJECT TO TAKE A LOOK AT THE INSIDE. REMEMBER TO PUT YOUR SECTION LINES ON A DIFFERENT LAYER I.E. "HATCH." |
| 62 | WITH THE POLE BRACKET YOU DRAW A FULL SECTION OF THE OBJECT.  REMEMBER THAT NO HIDDEN LINES WILL SHOW IN A SECTION VIEW. |
| 63 | T-BRACE GIVES YOU EXPERIENCE WITH REVOLVED SECTIONS.  BE SURE TO COPY THE PART YOU NEED FROM ONE OF YOUR OTHER VIEWS. |
| 64 | WITH THE CHISEL DRAWING YOU DRAW A REMOVED SECTION.  AGAIN BE SURE TO MAKE YOUR DRAWING AS EASY AS POSSIBLE BY USING YOUR EDITING COMMANDS. |
| 65 | ON THE HOLD COLLAR YOU DRAW A BROKEN SECTION.  USE THE POLYLINE TO CREATE YOUR CUTTING PLANE LINE. |
| 66 | WEB PULLEY LETS YOU DECIDE WHAT TYPE OF SECTION AND ORTHOGRAPHIC VIEW YOU WOULD USE TO CLEARLY DESCRIBE THE OBJECT. |
| 67 | SPOKE WHEEL GIVES YOU A CHANCE TO SECTION A SPOKE.  REMEMBER NOT TO HATCH TOO MUCH OF YOUR OBJECT. |
| 68 | VALVE BODY OFFERS YOU A FULL SECTION WITH SOME INTERESTING INSIDES.  FOR EXTRA CREDIT DO AN ISOMETRIC SECTION VIEW.  GOOD LUCK! |
| 69 | GIVEN AN ORTHOGRAPHIC VIEW OF A GAUGE COVER, DO A FULL SECTION.  WORK ON YOUR SPEED AND ACCURACY. REMEMBER TO USE YOUR EDITING COMMANDS.  YOU CAN DECREASE YOUR DRAWING TIME BY A FOURTH. |
| 70 | WITH THE SPINDLE BRACKET PERFORM THE INDICATED SECTION.  OMIT GEOMETRIC TOLERANCES IF NOT NEEDED. |

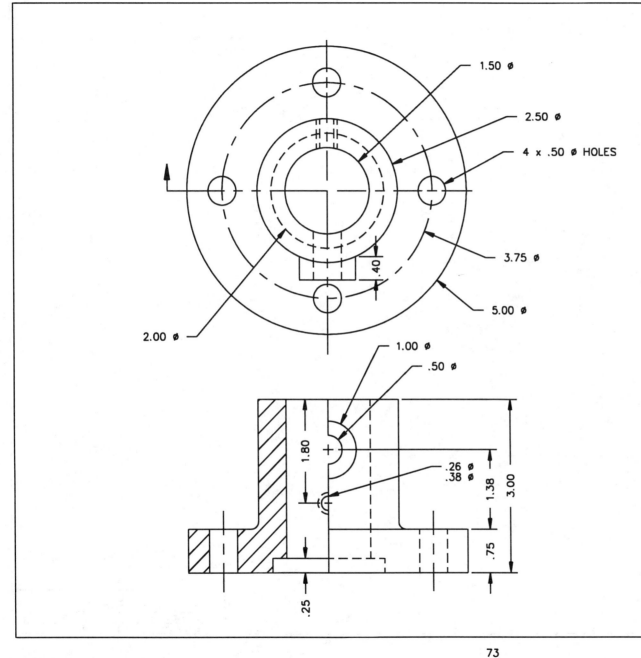

PROBLEM 61
HUB BASE

DRAW THE NECESSARY VIEWS AND THE INDICATED SECTION. INCLUDE ALL DIMENSIONS.
NOTE: ALL FILLETS ARE .125 R.

HINT:

CREATE A "HATCH" LAYER.
USE TRIM TO CREATE AREA TO HATCH.
USE FILLET TO REPLACE MISSING LINES.

| 61 | HUB BASE | 8 | 1020 |
|----|----------|---|------|
| NO. | PART | REQ | MATL |

USING AUTOCAD
HALF SECTION

MARK SIGL — SCALE: FULL
DATE: 6/1/91 — FIG. #: PRB 61

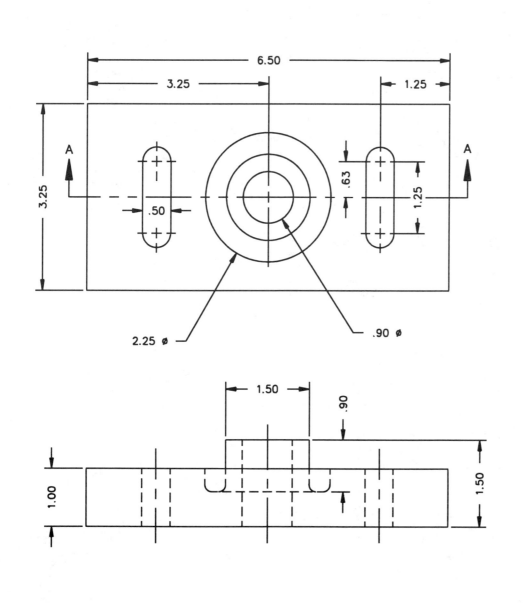

PROBLEM 62
POLE BRACKET

DRAW TOP VIEW AND A FULL
SECTION IN THE FRONT VIEW.
DO NOT FORGET TO DIMENSION.
NOTE: ALL FILLETS ARE .125 R.

| 62 | POLE BRACKET | 2 | 3050 |
|---|---|---|---|
| NO. | PART | REQ | MATL |

USING AUTOCAD
FULL SECTION

MARK SIGL · SCALE: FULL
DATE: 6/1/91 · FIG. #: PRB 62

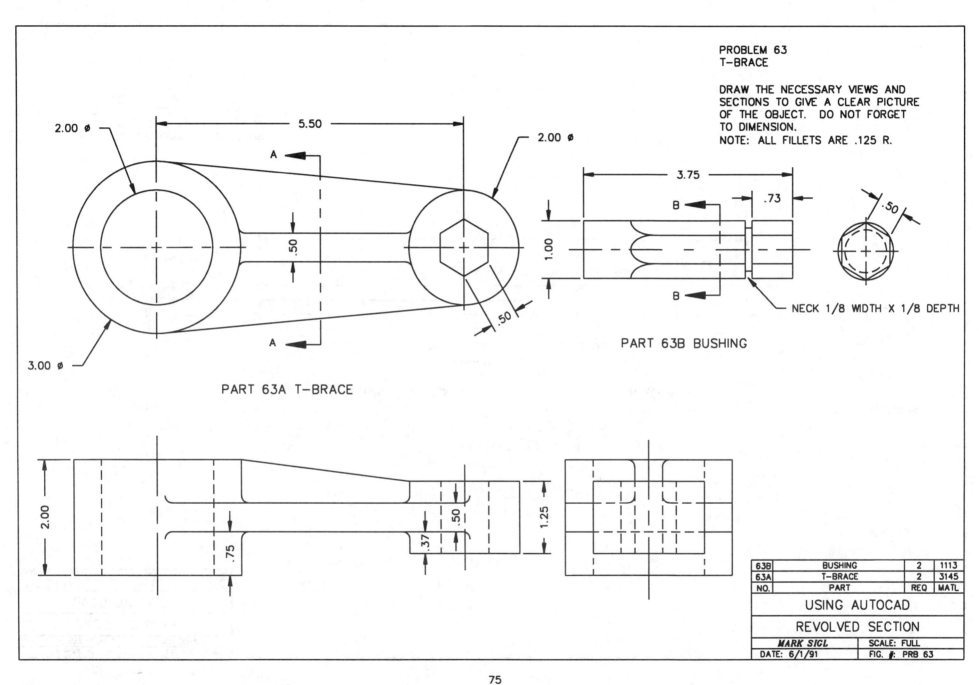

PROBLEM 64
CHISEL

DRAW OBJECT AND THE REMOVED
SECTIONS INDICATED. PLACE DIMENSIONS
IN APPROPRIATE LOCATION.
NOTE: AFTER SHARPING, HEAT TREAT.

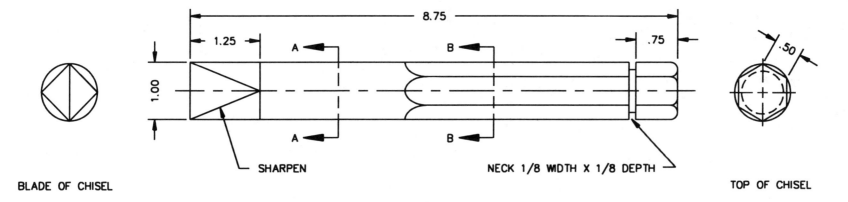

PART 64 CHISEL

| 64 | CHISEL | 1 | 4375 |
|----|--------|---|------|
| NO. | PART | REQ | MATL |

USING AUTOCAD
REMOVED SECTION

MARK SICL — SCALE: FULL
DATE: 6/1/91 — FIG. #: PRB 64

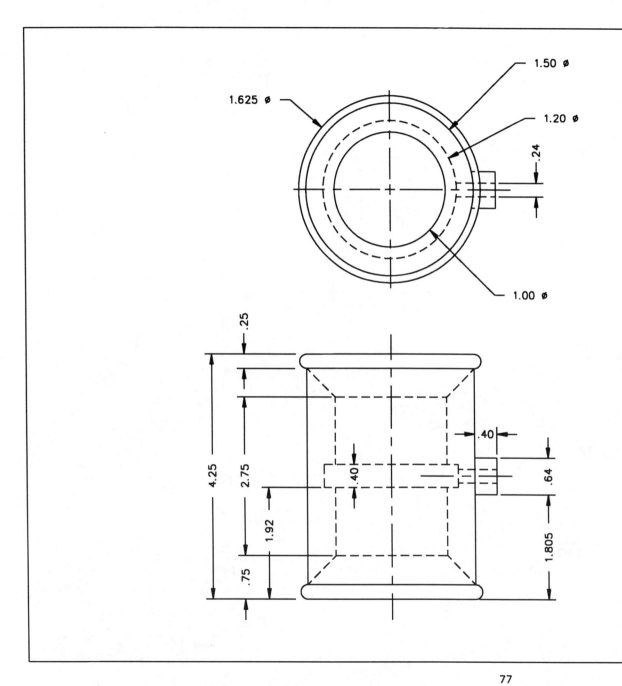

PROBLEM 65
HOLD COLLAR

DRAW NECESSARY VIEWS AND
A BROKEN SECTION VIEW.

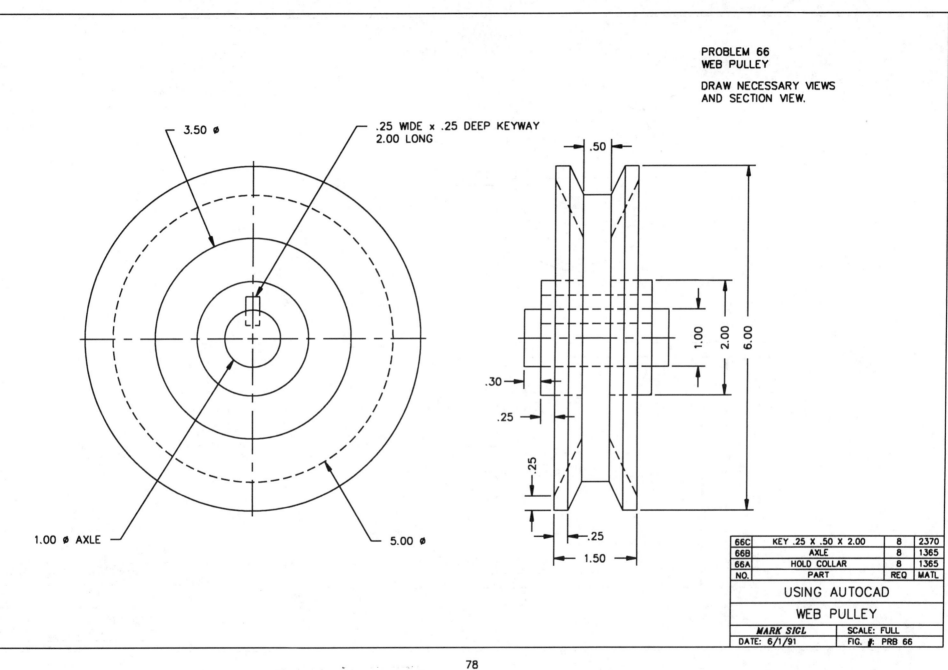

PROBLEM #67
SPOKE WHEEL

DRAW GIVEN VIEWS,
DIMENSION, AND SECTION.
NOTE: FILLETS ARE .125 R.

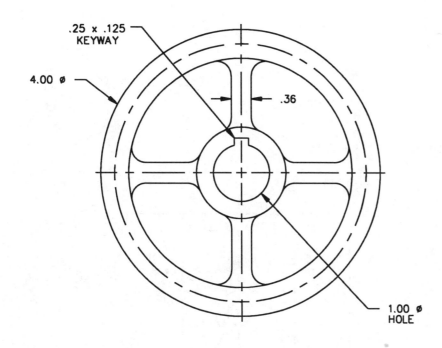
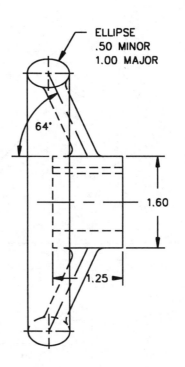

| 67 | SPOKE WHEEL | 1 | 2350 |
|---|---|---|---|
| NO. | PART | REQ | MATL |

USING AUTOCAD

SPOKE WHEEL

| MARK SIGL | SCALE: FULL |
|---|---|
| DATE: 6/1/91 | FIG. #: PRB 67 |

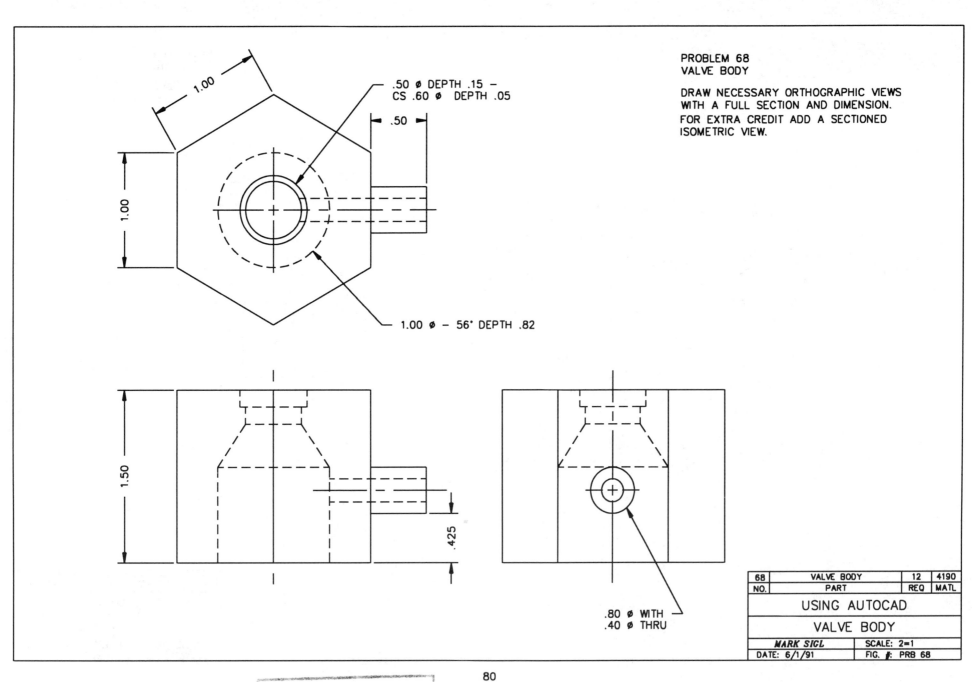

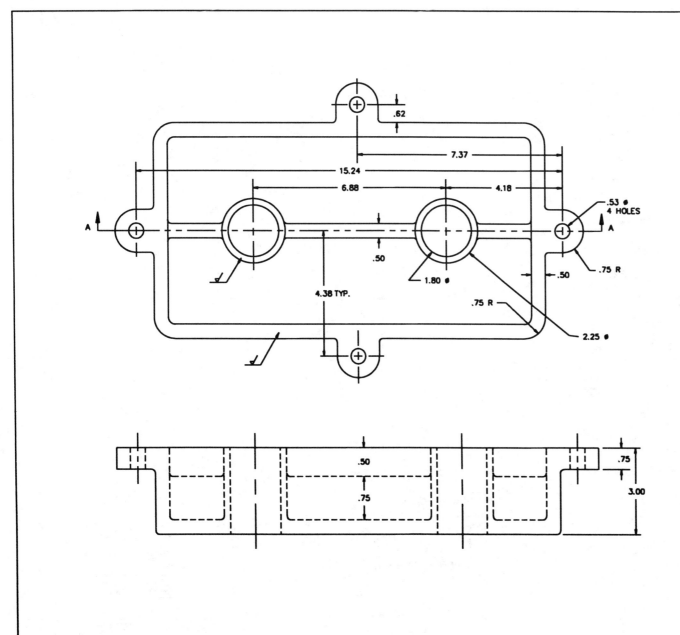

PROBLEM 69
GAUGE COVER

DRAW NECESSARY VIEWS
AND DIMENSIONS.

NOTES:

1. ALL FILLETS ARE .125 R.
2. MACHINED SURFACES ARE INDICATED.
3. MATERIAL IS SAE 1020.
4. TOLERANCE IS ± .01.

HINT:

USE EDITING COMMANDS
TO CUT DRAWING TIME
BY A FOURTH.

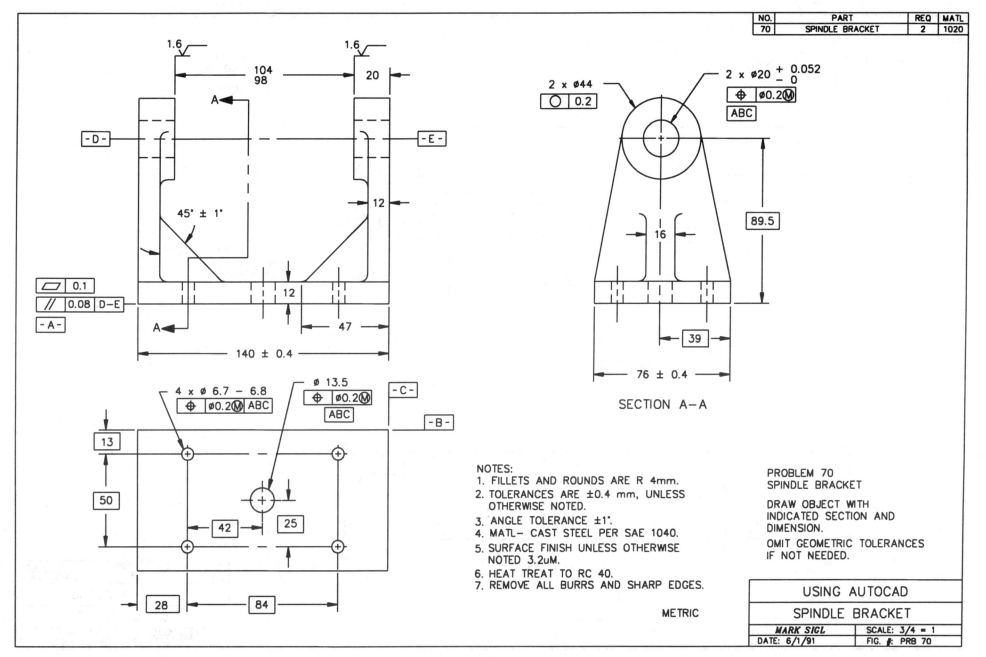

# CHAPTER 8
# AUXILIARY & INQUIRY

THIS CHAPTER EXPOSES YOU TO PERFORMING AUXILIARY VIEWS AND USING INQUIRY COMMANDS TO AID YOU WITH DRAWING.
AUXILIARY VIEWS ARE USED TO OBTAIN THE TRUE LENGTH OR SHAPE OF AN OBJECT THAT IS NOT PARALLEL TO THE REGULAR
PLANES OF PROJECTION.  YOU NEED TO PERFORM AUXILIARY VIEWS TO OBTAIN THE NECESSARY INFORMATION.
BE SURE TO USE YOUR EDITING COMMANDS TO HELP YOU DRAW THE OBJECTS FASTER AND MORE ACCURATELY.

PROBLEM                                                        TOPIC

71      EXERCISE EXPOSES YOU TO THE TRUE MEANING OF A TRUE LENGTH LINE.  THE USE OF THE OFFSET COMMAND WILL
        MAKE YOUR LIFE EASIER AND ENABLE YOU TO DRAW THE OBJECT FASTER.

72      FIND SLOPE OF A LINE, BEARING (BOTH COMPASS AND AZIMUTH), AND THE PERCENT GRADE.

73      FINE TRUE SLOPE OF AN OBLIQUE LINE.

74      FIND A PLANE IN ITS TRUE SIZE AND SHAPE.

75      DETERMINE PIERCING POINT OF A LINE AND A PLANE.

76      USE A SECONDARY AUXILIARY IN OBTAINING THE NECESSARY INFORMATION SUCH THE PERIMETER AND AREA OF AN OBLIQUE PLANE.

77      DRAW AN AUXILIARY VIEW OF A GAUGE BLOCK TO SHOW ACTUAL SIZE AND SHAPE.

78      USE AUXILIARY VIEWS TO HELP DETERMINE THE AREA OF THE SLOPED SURFACES AND TO CALCULATE THE AREA
        OF THE SLOTS TO A NET SURFACE AREA.  FOR EXTRA CREDIT DETERMINE THE VOLUME OF THE ANGLE PLATE.

79      LOCATE DRILL HOLES ON THE TRUE SIZE AND SHAPE OF THE ANGLE BRACKET.  YOU WOULD HATE TO GIVE FORESHORTENED
        DIMENSIONS.

80      OBTAIN TRUE SIZE AND SHAPE OF A CONNECTING BAR.

PROBLEM 71
TRUE LENGTH OF A LINE

FIND TRUE LENGTH OF $\overline{AB}$ TO NEAREST THOUSANDTH.

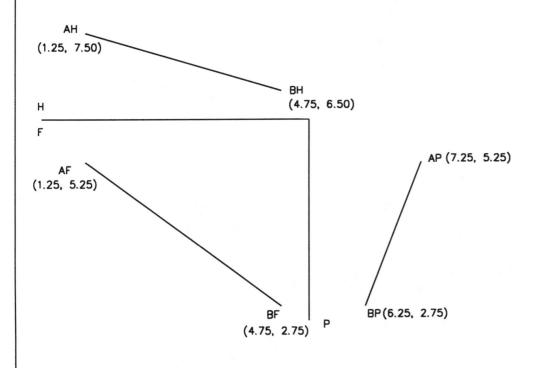

| USING AUTOCAD | |
|---|---|
| TRUE LENGTH OF A LINE | |
| *MARK SIGL* | SCALE: FULL |
| DATE: 6/1/91 | FIG. #: PRB 71 |

PROBLEM 72
BEARING OF A LINE

USE PROBLEM 71 (PROTOTYPE) TO DO THIS EXERCISE (PRB72=PRB71).
FIND SLOPE, PERCENT GRADE, AND BEARING (BOTH COMPASS AND AZIMUTH)
OF $\overline{AB}$ TO NEAREST THOUSANDTH.

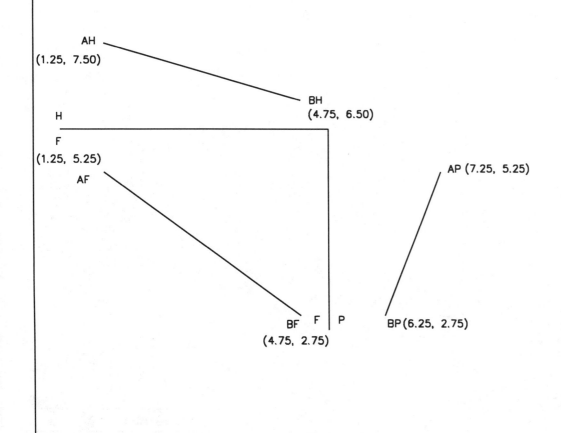

| USING AUTOCAD |||
|---|---|---|
| BEARING OF A LINE |||
| *MARK SIGL* || SCALE: FULL |
| DATE: 6/1/91 || FIG. #: PRB 72 |

PROBLEM 73
TRUE SLOPE OF A LINE

FIND TRUE LENGTH AND SLOPE OF $\overline{AB}$ TO NEAREST THOUSANDTH.
LABEL END POINTS IN AUXILIARY VIEW.
FIND LINE IN TOP VIEW AND LABEL ITS END POINTS.

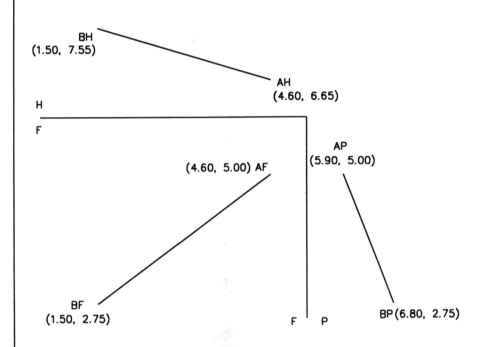

| USING AUTOCAD |
|---|
| SLOPE OF A LINE |
| MARK SIGL | SCALE: FULL |
| DATE: 6/1/91 | FIG. #: PRB 73 |

PROBLEM 74
TRUE SIZE AND SHAPE OF A PLANE

FINE AREA AND PERIMETER OF THE PLANE
TO NEAREST THOUSANDTH.

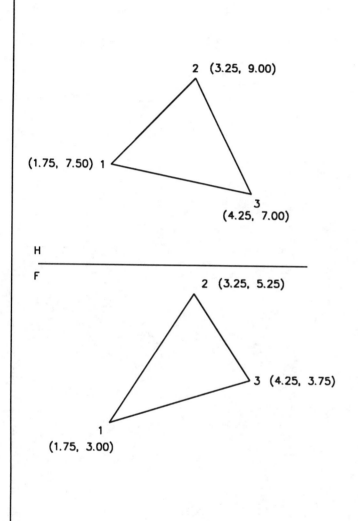

| USING AUTOCAD |||
|---|---|---|
| TRUE SHAPE OF PLANE |||
| **MARK SIGL** | SCALE: FULL ||
| DATE: 6/1/91 | FIG. #: PRB 74 ||

PROBLEM 75
PIERCING POINT BY AUXILIARY VIEW

FIND THE PIERCING POINT OF $\overline{AB}$ AND △ 123. ALSO
FIND ANGLE BETWEEN LINE AND PLANE.

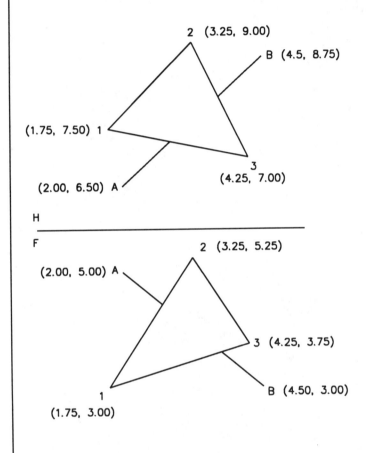

| USING AUTOCAD |||
|---|---|---|
| PIERCING POINT |||
| *MARK SIGL* | SCALE: | FULL |
| DATE: 6/1/91 | FIG. #: | PRB 75 |

PROBLEM 76
SECONDARY AUXILIARY VIEW

FIND TRUE SHAPE OF OBLIQUE SURFACE
AND DETERMINE AREA AND PERIMETER TO NEAREST
THOUSANDTH. DO NOT FORGET TO DIMENSION.

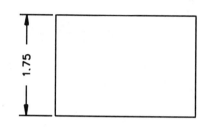

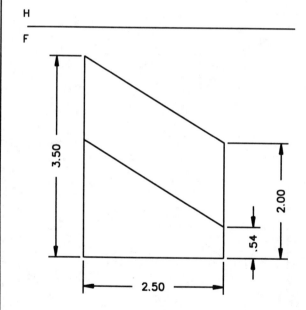

| USING AUTOCAD |||
|---|---|---|
| SECONDARY AUXILIARY VIEW |||
| *MARK SIGL* || SCALE: FULL |
| DATE: 6/1/91 || FIG. #: PRB 76 |

PROBLEM 77
GAUGE BLOCK

DRAW ORTHOGRAPHIC VIEWS AND DIMENSION.
CREATE AUXILIARY VIEW OF SLOPED
SURFACE.

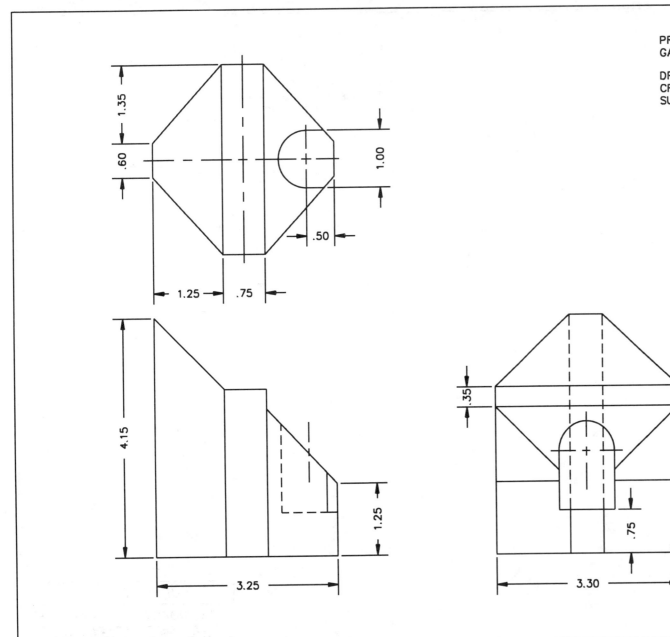

| USING AUTOCAD |||
|---|---|---|
| GAUGE BLOCK |||
| MARK SIGL | SCALE: FULL ||
| DATE: 6/1/91 | FIG. #: PRB 77 ||

PROBLEM 78
ANGLE PLATE

DRAW OBJECT, AUXILIARY VIEWS, AND DIMENSION. DETERMINE PERIMETER AND SURFACE AREAS TO NEAREST THOUSANDTH. FIND GROSS AREA (INCLUDING SLOT), SLOT AREA (AREA OF HOLES), AND NET AREA (GROSS − SLOT = NET). FOR EXTRA CREDIT DETERMINE VOLUME OF PLATE.

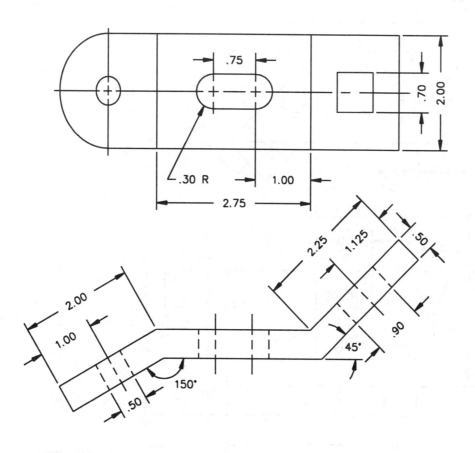

| USING AUTOCAD |||
|---|---|---|
| ANGLE PLATE |||
| *MARK SIGL* | SCALE: | FULL |
| DATE: 6/1/91 | FIG. #: | PRB 78 |

PROBLEM 79
ANGLE BRACKET

DRAW ORTHOGRAPHIC VIEWS, AUXILIARY VIEWS, AND DIMENSION.
FOR EXTRA CREDIT FIND NET AREA, GROSS AREA, AND
SLOT AREA TO NEAREST THOUSANDTH.

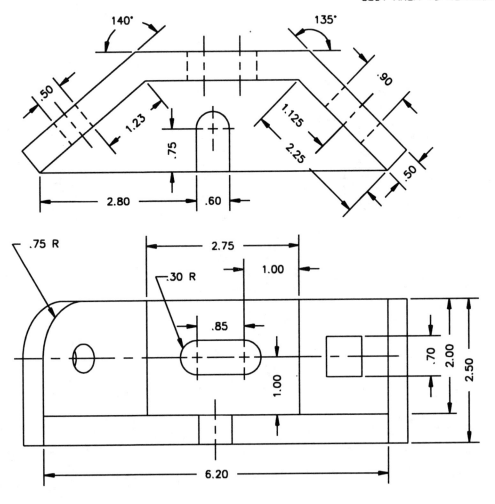

| USING AUTOCAD |||
|---|---|---|
| ANGLE BRACKET |||
| MARK SIGL | SCALE: FULL ||
| DATE: 6/1/91 | FIG. #: PRB 79 ||

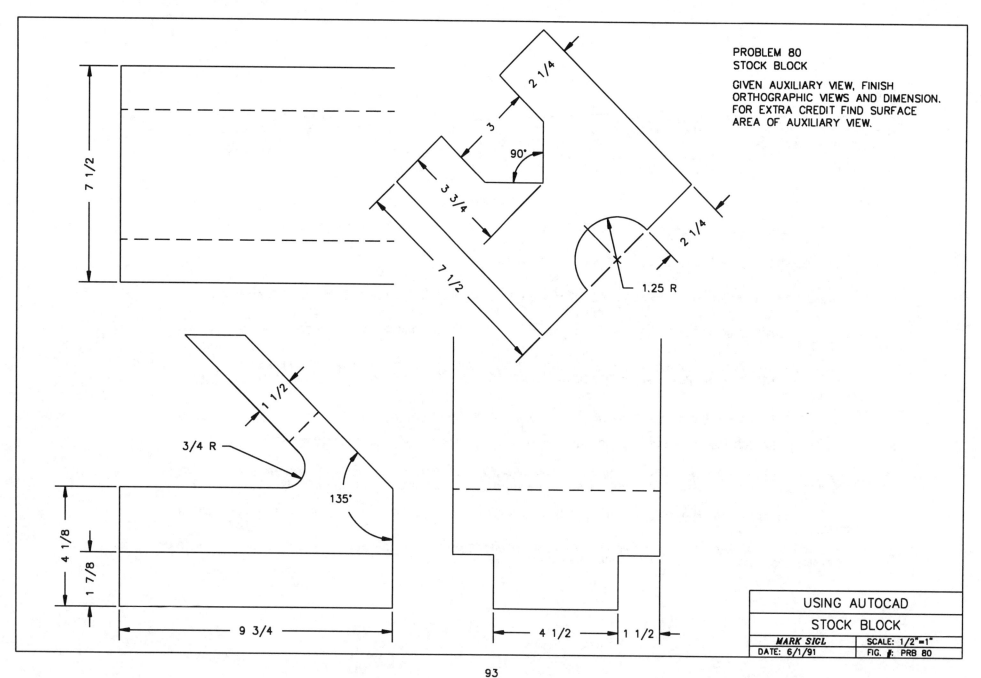

# CHAPTER 9
# DEVELOPMENTS

THIS CHAPTER GIVES YOU EXPERIENCE USING YOUR EDITING COMMANDS IN DESCRIPTIVE GEOMETRY AND IN CREATING DEVELOPMENTS.
THESE EXERCISES INCREASE YOUR DRAWING ACCURACY BECAUSE PRECISION IS NECESSARY TO ASSEMBLE THE DEVELOPMENTS.
YOW ALSO EXPERIENCE HOW DEVELOPMENTS ARE LAID OUT AND ASSEMBLED IN THE PACKING AND SHEET METAL INDUSTRY.
THE COMPUTER ALLOWS YOU TO BE EXTREMELY ACCURATE AND INCREASES YOUR SPEED WITH COMMANDS LIKE COPY, MIRROR,
DIVIDE, AND OFFSET.  YOU WILL FIND THIS CHAPTER A CHALLENGING ONE.

PROBLEM                                                        TOPIC

81     CREATE A DEVELOPMENT OF A BOX.  THIS EXERCISE TESTS YOUR USE OF EDITING COMMANDS AND DRAWING ACCURACY.

82     THE DEVELOPMENT OF A "CONE" IS USED TO INCREASE YOUR LAYOUT SKILLS.

83     PYRAMID POWER IS IN THE DEVELOPMENT HERE.  REMEMBER TO LET THE EDITING COMMANDS DO ALL THE WORK FOR YOU.

84     USE OF DESCRIPTIVE GEOMETRY PLAYS MANY ROLES IN THE DRAFTING FIELD.  THIS EXERCISE EXPOSES YOU TO
       SOME BASICS.

85     A MORE INDEPTH PROBLEM DEALING WITH A LINE INTERSECTING A PLANE.  FIND THE AREA OF A PLANE AND THE TRUE LENGTH OF
       THE LINE INTERSECTING THE PLANE AND INTERSECTION ANGLE.  THIS ONE PREPARES YOU FOR THE REAL WORLD OR AT
       LEAST PROBLEM 90.

86     THE DEVELOPMENT OF A TRUNCATED CYLINDER OR AT LEAST A NEW DESIGN FOR A SODA CAN.

87     TRUNCATED HEXAGON OFFERS A MODULAR APPROACH TO HOUSING DESIGN OR THE ROOF DESIGN INDUSTRY.

88     PIPING AND DUCTING INDUSTRY IS THE SETTING FOR THIS PROBLEM.  EXERCISE SHOWS YOU HOW TO MANAGE
       THE INTERSECTION OF CORNERS.  FOR THOSE WHO HAVE MASTERED THE EDITING COMMANDS, YOUR WORK IS HALF DONE!

89     HERE IS A PROBLEM WITH A DIFFERENT ANGLE.  IT IS A TRUNCATED CONE.  ELLIPSES PLAY A BIG PART IN DEALING WITH
       CIRCLES AT AN ANGLE.

90     YOU ARE NOW READY FOR THE "TRANSITION": TURNING A RECTANGLE INTO A CIRCLE.  YOU WILL NEED TO USE ALL OF YOUR EDITING
       COMMANDS AND LAY OUT SKILLS TO GET YOU THROUGH THIS ONE.  GOOD LUCK !

PROBLEM 81
BOX DEVELOPMENT

SET UP THE ORTHOGRAPHIC VIEW OF THE BOX AND PROCEED TO DEVELOP THE BOX WITH .25" GLUE TABS AND FOLDING LINES. BE SURE TO USE OFFSET AND REFERENCE TO OBTAIN CORRECT LENGTHS. LET THE DRAWING PROVIDE YOU WITH THE MEASUREMENTS.

IF YOU HAVE TROUBLE FOLLOWING THE UNFOLDING WITH THE MEASUREMENTS, THEN USE A STRETCHOUT LINE AND LABEL YOUR LENGTHS, WIDTHS, AND HEIGHTS WITH LETTERS AND NUMBERS. THE PROCEDURE IS THE SAME AS THE MANUAL METHOD. THE COMPUTER LETS YOU BE EXACT WITH THE MEASUREMENTS AND PRODUCES NEAT UNIFORM DRAWINGS AND ALLOWS YOU TO TAKE SHORT CUTS.

TO CHECK FOR ACCURACY, CUT OUT THE DEVELOPMENT AND ASSEMBLE.

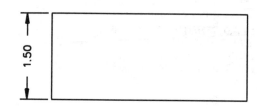

TOP VIEW

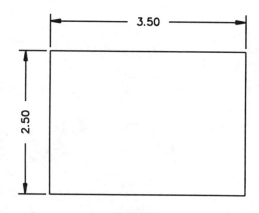

FRONT VIEW

| USING AUTOCAD ||
|---|---|
| BOX DEVELOPMENT ||
| MARK SIGL | SCALE: FULL |
| DATE: 6/1/91 | FIG. #: PRB 81 |

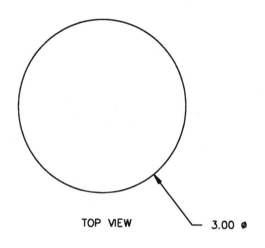

TOP VIEW — 3.00 ⌀

PROBLEM 82
CONE DEVELOPMENT

SET UP THE ORTHOGRAPHIC VIEW OF THE CONE AND PROCEED DEVELOP
THE CONE WITH .25" GLUE TABS AND FOLDING LINES. USE LIST, ARRAY, AND MEASURE TO
OBTAIN CORRECT LENGTHS. LET THE DRAWING PROVIDE YOU WITH THE MEASUREMENTS.

IF YOU HAVE TROUBLE FOLLOWING THE UNFOLDING WITH THE MEASUREMENTS, THEN USE A
STRETCHOUT LINE AND LABEL YOUR LENGTHS, WIDTHS, AND HEIGHTS WITH LETTERS AND NUMBERS.
THE PROCEDURE IS THE SAME AS THE MANUAL METHOD. THE COMPUTER LETS YOU BE
EXACT WITH THE MEASUREMENTS AND PRODUCES NEAT UNIFORM DRAWINGS AND ALLOWS YOU TO
TAKE SHORT CUTS.

TO CHECK FOR ACCURACY, CUT OUT THE DEVELOPMENT AND ASSEMBLE.

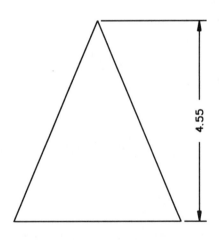

4.55

FRONT VIEW

| USING AUTOCAD |||
|---|---|---|
| CONE DEVELOPMENT |||
| *MARK SIGL* | SCALE: FULL ||
| DATE: 6/1/91 | FIG. #: PRB 82 ||

TOP VIEW

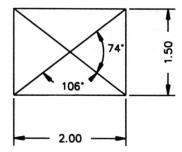

PROBLEM 83
PYRAMID DEVELOPMENT

SET UP THE ORTHOGRAPHIC VIEW OF THE PYRAMID AND PROCEED TO DEVELOP
THE PYRAMID WITH .25" GLUE TABS AND FOLDING LINES.  BE SURE TO USE OFFSET, COPY,
CIRCLE, AND MIRROR.  LET THE DRAWING PROVIDE YOU WITH THE MEASUREMENTS.
REMEMBER YOU NEED TO FIRST CREATE AN AUXILIARY VIEW TO OBTAIN THE HEIGHT.

TO CHECK FOR ACCURACY, CUT OUT THE DEVELOPMENT AND ASSEMBLE.

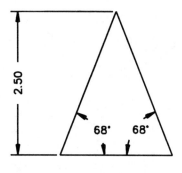

FRONT VIEW

| USING AUTOCAD |||
|---|---|---|
| PYRAMID DEVELOPMENT |||
| *MARK SIGL* || SCALE: FULL |
| DATE: 6/1/91 || FIG. #: PRB 83 |

PROBLEM 84
INTERSECTIONS OF TWO LINES

FIND:
1. SHORTEST DISTANCE BETWEEN SKEWED LINES.
2. TRUE LENGTH OF $\overline{CD}$.
3. ANGLE BETWEEN $\overline{AB}$ AND $\overline{CD}$.
4. STATE ANSWERS TO NEAREST TEN THOUSANDTH.

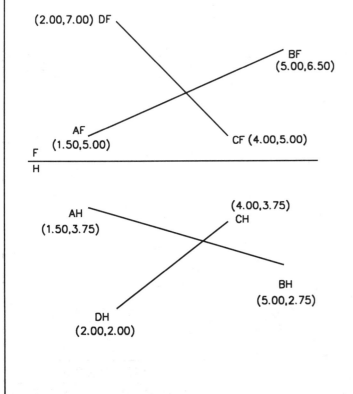

| USING AUTOCAD |||
|---|---|---|
| INTERSECTIONS |||
| MARK SIGL | SCALE: FULL ||
| DATE: 6/1/91 | FIG. #: PRB 84 ||

PROBLEM 85
INTERSECTION OF A LINE AND PLANE

SET UP THE ORTHOGRAPHIC VIEW OF THE LINE AND PLANE AND PROCEED TO CREATE
THE LINE AND PLANE.  BE SURE TO USE EXTEND, TRIM, LIST, OFFSET, AND REFERENCE TO
OBTAIN CORRECT LENGTHS.  LET THE DRAWING PROVIDE YOU WITH THE MEASUREMENTS.
FIND AREA OF THE PLANE, PERIMETER OF THE PLANE, ANGLE BETWEEN LINE AND THE
PLANE, AND TRUE LENGTH OF THE LINE.  GOOD LUCK!

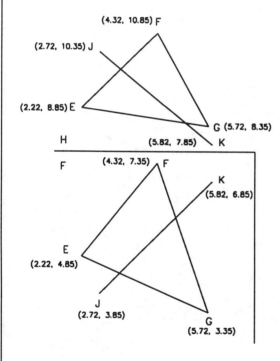

USING AUTOCAD
INTERSECTION
MARK SIGL | SCALE: FULL
DATE: 6/1/91 | FIG. #: PRB 85

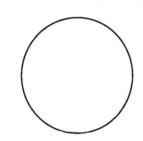

PROBLEM 86
TRUNCATED CYLINDER DEVELOPMENT

SET UP THE ORTHOGRAPHIC VIEW OF THE CYLINDER AND PROCEED TO DEVELOP
THE CYLINDER WITH .25" GLUE TABS AND FOLDING LINES. BE SURE TO USE OFFSET, COPY, CIRCLE,
AND MIRROR TO CHECK FOR ACCURACY, CUT OUT AND ASSEMBLE THE DEVELOPMENT.

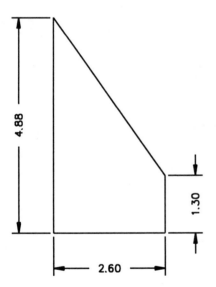

| USING AUTOCAD | |
|---|---|
| CYLINDER | |
| MARK SIGL | SCALE: FULL |
| DATE: 6/1/91 | FIG. #: PRB 86 |

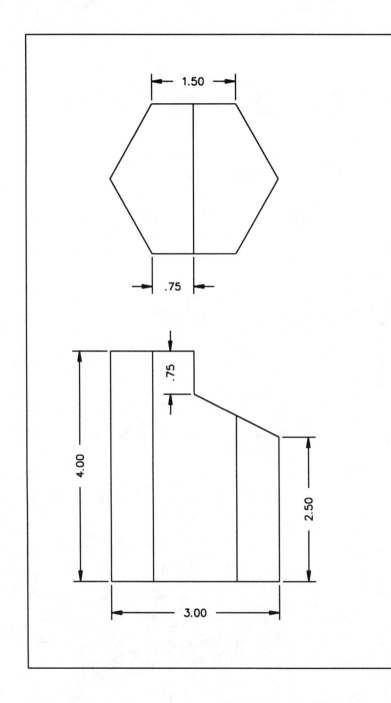

PROBLEM 87
TRUNCATED HEXAGON

CONSTRUCT TRUNCATED HEXAGON.
BE SURE TO INCLUDE GLUE TABS.
PLOT AND CUT OUT TO
CHECK FOR ACCURACY.

| USING AUTOCAD |||
|---|---|---|
| TRUNCATED HEXAGON |||
| *MARK SIGL* | SCALE: FULL ||
| DATE: 6/1/91 | FIG. #: PRB 87 ||

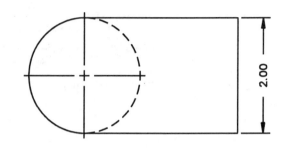

PROBLEM 88
TWO-PIECE ELBOW DEVELOPMENT

SET UP ORTHOGRAPHIC VIEW OF CYLINDER AND PROCEED TO DEVELOP THE CYLINDER WITH .25" GLUE TABS AND FOLDING LINES. BE SURE TO USE OFFSET, COPY, CIRCLE, AND MIRROR.

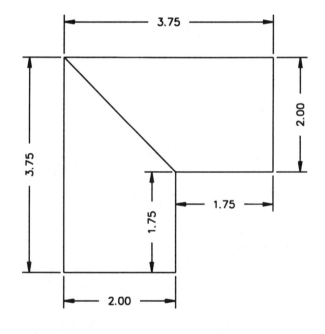

| USING AUTOCAD |||
|---|---|---|
| TWO-PIECE ELBOW |||
| MARK SIGL | SCALE: FULL ||
| DATE: 6/1/91 | FIG. #: PRB 88 ||

**PROBLEM 89**
**TRUNCATED CONE**

CONSTRUCT TRUNCATED CONE. BE SURE TO INCLUDE GLUE TABS AND FOLD LINES. PLOT AND CUT OUT TO CHECK FOR ACCURACY.

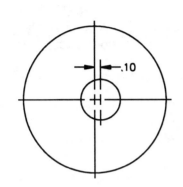

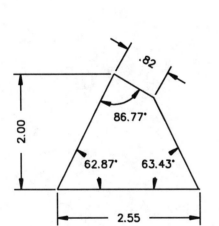

| USING AUTOCAD |||
|---|---|---|
| TRUNCATED CONE |||
| *MARK SIGL* | SCALE: FULL ||
| DATE: 6/1/91 | FIG. #: PRB 89 ||

103

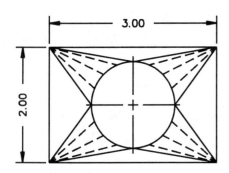

PROBLEM 90
RECTANGLE TO CIRCLE

SET UP THE ORTHOGRAPHIC VIEW OF THE OBJECT AND PROCEED TO CREATE THE OBJECT WITH .25" GLUE TABS AND FOLDING LINES. BE SURE TO USE ARRAY, DIVIDE, COPY, CIRCLE, AND MIRROR. ADD TOP AND BOTTOM TO PRODUCE A CONTAINER.

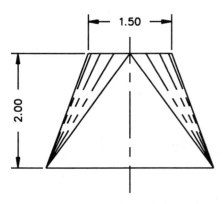

| USING AUTOCAD |||
|---|---|---|
| TRANSITION |||
| MARK SIGL | SCALE: FULL ||
| DATE: 6/1/91 | FIG. #: PRB 90 ||

# CHAPTER 10
# 3-DIMENSIONAL MODELS

THIS CHAPTER COVERS CREATING 3-DIMENSIONAL DRAWINGS.  WITH COMPUTERS AIDING ENGINEERS, GRAPHIC ILLUSTRATORS, AND MANUFACTURERS, IT HAS BECOME NECESSARY TO CREATE 3-D MODEL DRAWINGS TO BE USED FOR STRUCTURE TESTING, COMPUTER AIDED MANUFACTURING (CAM), RENDERINGS, AND ADVERTISEMENT.  IT IS NECESSARY TO VIEW OBJECTS IN 3-D INSTEAD OF 2-D FOR SPACIAL PLANNING AND FABRICATION OF OBJECTS.  BE SURE TO USE CHANGE (ELEVATION AND THICKNESS), 3DLINE, 3DFACE, REVSURF, RULESURF, TABSURF, 3-D OBJECTS, EDGESURF, 3DMESH, SURFTAB1, SURFTAB2, AND HIDE.  DO NOT FORGET ABOUT THE OTHER COMMANDS TO HELP YOU SUCH AS VPOINT, VPORTS, UCS, MVIEW, MSPACE, PSPACE, PFACE, UCSICON , AND DVIEW.

| PROBLEM | TOPIC |
|---|---|
| 91 | VISE BLOCK IS AN EXCELLENT EXAMPLE OF A PART THAT CAN BE MANUFACTURED WITH A CAM SYSTEM. |
| 92 | T-BLOCK IS CONSTRUCTED USING THE CHANGE OR TABSURF COMMANDS.  USE YOUR VPORTS AND VPOINT TO HELP YOU VISUALIZE THE DRAWING. |
| 93 | WITH THE BEARING BASE YOU ARE EXPOSED TO 3-D ARCS AND CIRCLES.  FOR EXTRA CREDIT SECTION THE OBJECT.  THIS FAMILIARIZES YOU WITH THE USER COORDINATE SYSTEM (UCS). |
| 94 | POST BRACKET EXPOSES YOU TO DOING A 3-D SECTION DRAWING.  REMEMBER TO USE THE UCS FOR HATCHING YOUR SECTION. |
| 95 | 3-D MODEL OF A SPOCKET IS AN EXAMPLE OF AN ITEM THAT CAN BE MILLED WITH A CAM SYSTEM. |
| 96 | IMAGINE DESIGNING YOUR OWN WINE GLASS AT LEAST BUILDING A 3-D DRAWING. |
| 97 | ROCKER ARM OFFERS A CHALLENGE WITH A SLOT CUT IN THE OBJECT AT A DIFFERENT ANGLE THAN THE HOLES. |
| 98 | WITH THE SLIDE STOP YOU DEAL WITH INCLINED LINES IN 3-D.  REMEMBER TO USE 3DLINE AND 3DFACE. |
| 99 | STEP-V PULLEY OFFERS A LOT OF RECESSED SURFACES.  AGAIN THIS SHOULD AT BE A PROBLEM TO A SKILLED CAD USER WHO COULD CREATE THIS OBJECT IN A FEW SHORT STEPS (i.e.,. BLOCK/COPY, MIRROR, AND REVOLVED). |
| 100 | FINAL CHALLENGE IS TO DRAW THE FOUR COMPONETS IN 3-D AND ASSEMBLE THE PARTS. |

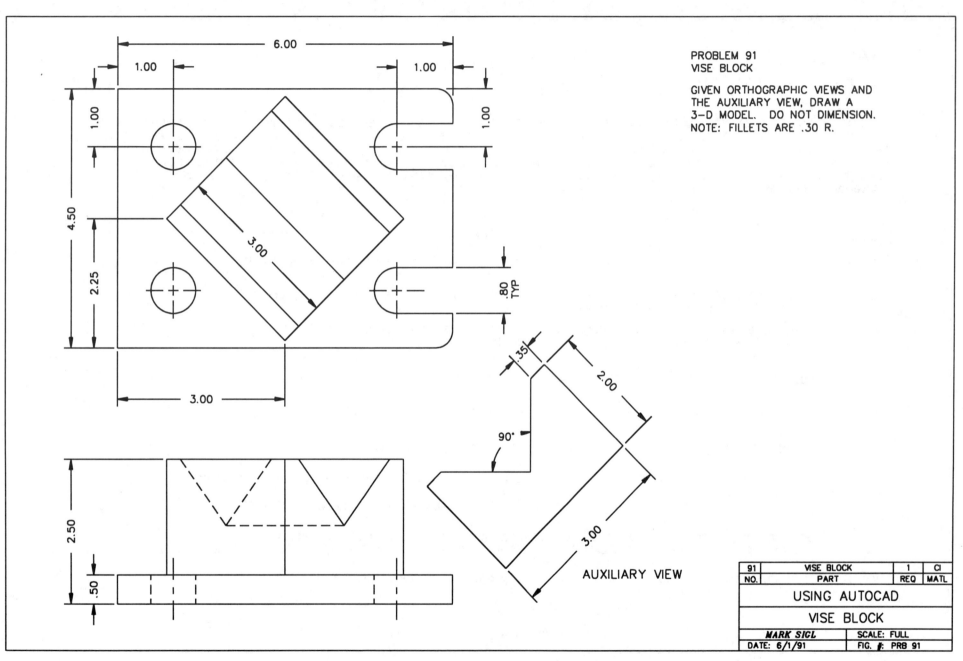

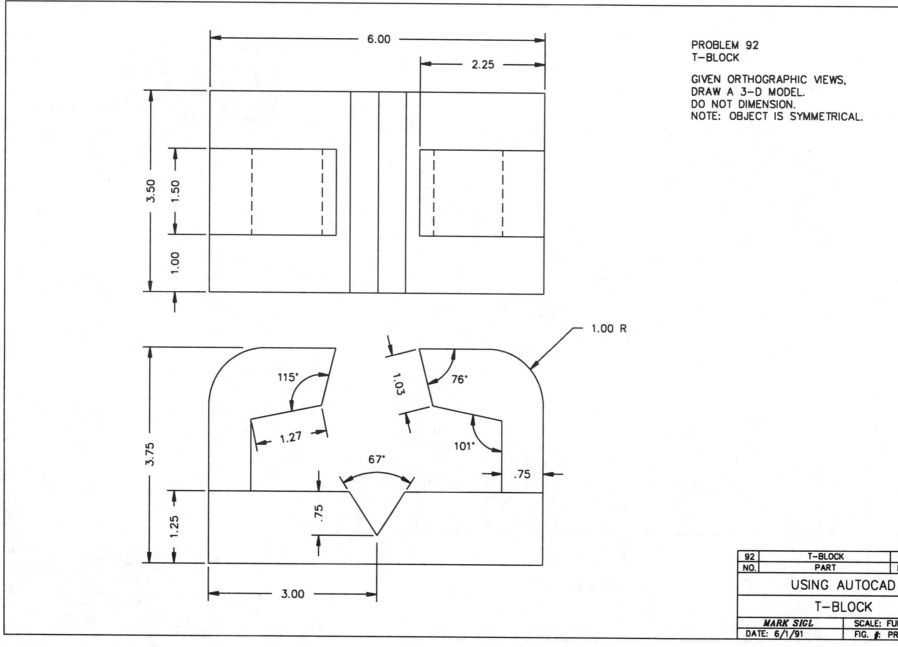

PROBLEM 92
T-BLOCK

GIVEN ORTHOGRAPHIC VIEWS,
DRAW A 3-D MODEL.
DO NOT DIMENSION.
NOTE: OBJECT IS SYMMETRICAL.

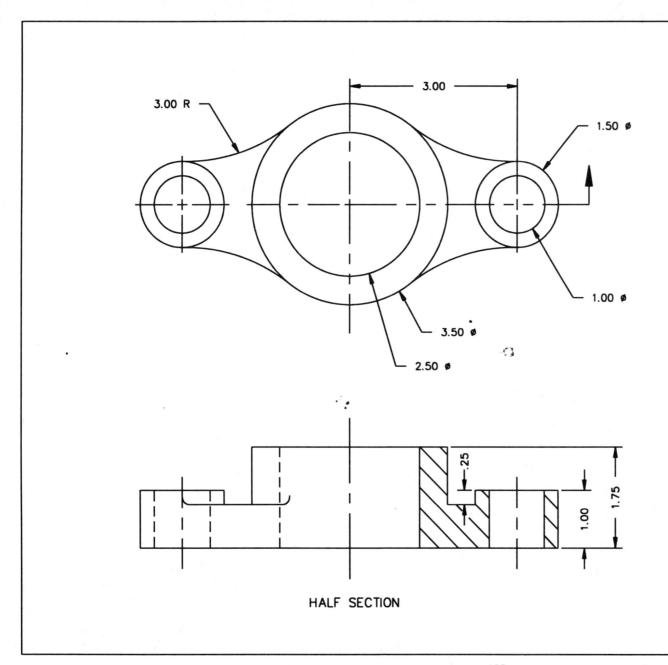

PROBLEM 93
BEARING BASE

GIVEN ORTHOGRAPHIC VIEWS,
DRAW A 3-D MODEL.
FOR EXTRA CREDIT DO A HALF OR A FULL
SECTION 3-D MODEL AND ADD SECTION LINES.
DO NOT DIMENSION.

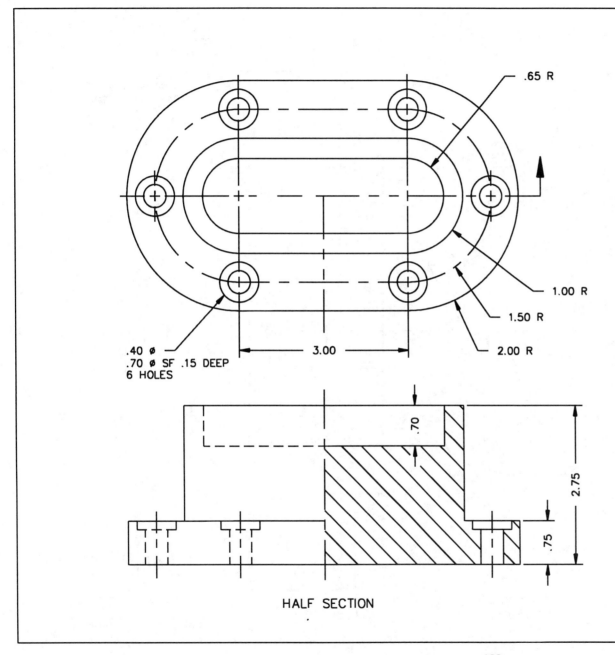

PROBLEM 94
POST BRACKET

USE ORTHOGRAPHIC VIEWS AND
HALF-SECTION VIEW TO DRAW A 3-D MODEL.
FOR EXTRA CREDIT DO A HALF OR A FULL
SECTION 3-D MODEL AND ADD SECTION LINES.
DO NOT DIMENSION.

| 94 | POST BRACKET | 24 | 1155 |
|---|---|---|---|
| NO. | PART | REQ | MATL |

USING AUTOCAD
POST BRACKET

| *MARK SIGL* | SCALE: FULL |
|---|---|
| DATE: 6/1/91 | FIG. #: PRB 94 |

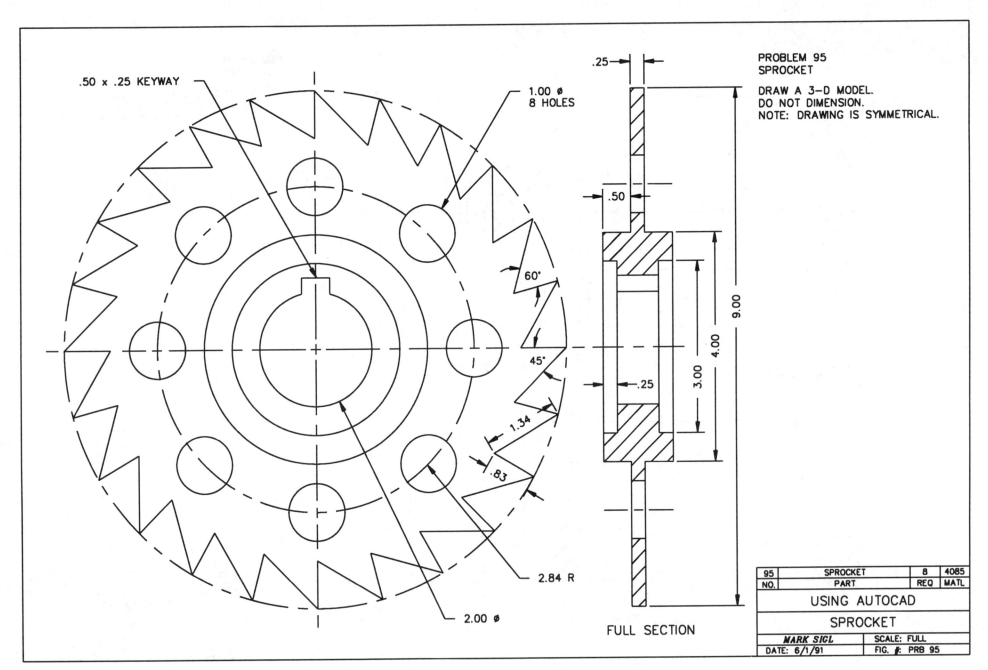

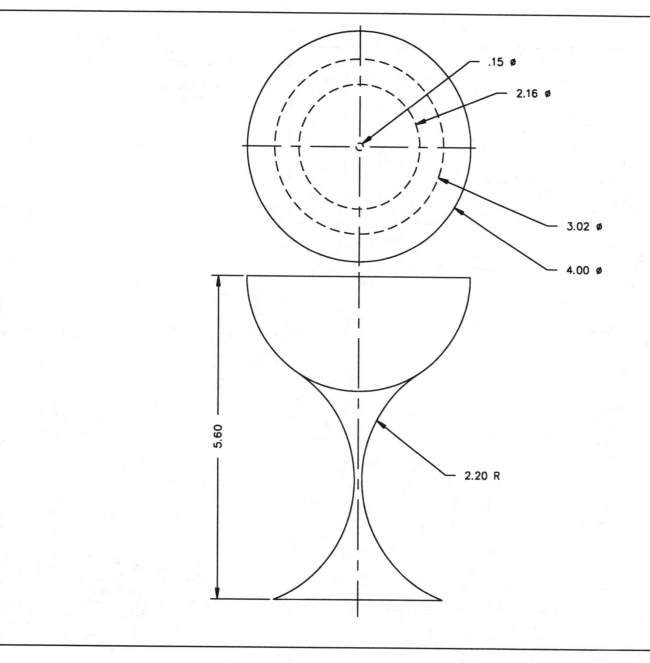

PROBLEM 96
STOP BUTTON

GIVEN ORTHOGRAPHIC VIEWS,
DRAW A 3-D MODEL.
DO NOT DIMENSION.
PLOT OUT THE TWO ORTHOGRAPHIC
VIEWS PLUS A 3-D VIEW ON THE
SAME DRAWING.
NOTE: DRAWING IS SYMMETRICAL.

| 96 | WINE GLASS | 12 | CRYTL |
|---|---|---|---|
| NO. | PART | REQ | MATL |

USING AUTOCAD

WINE GLASS

| MARK SICL | SCALE: FULL |
|---|---|
| DATE: 6/1/91 | FIG. #: PRB 96 |

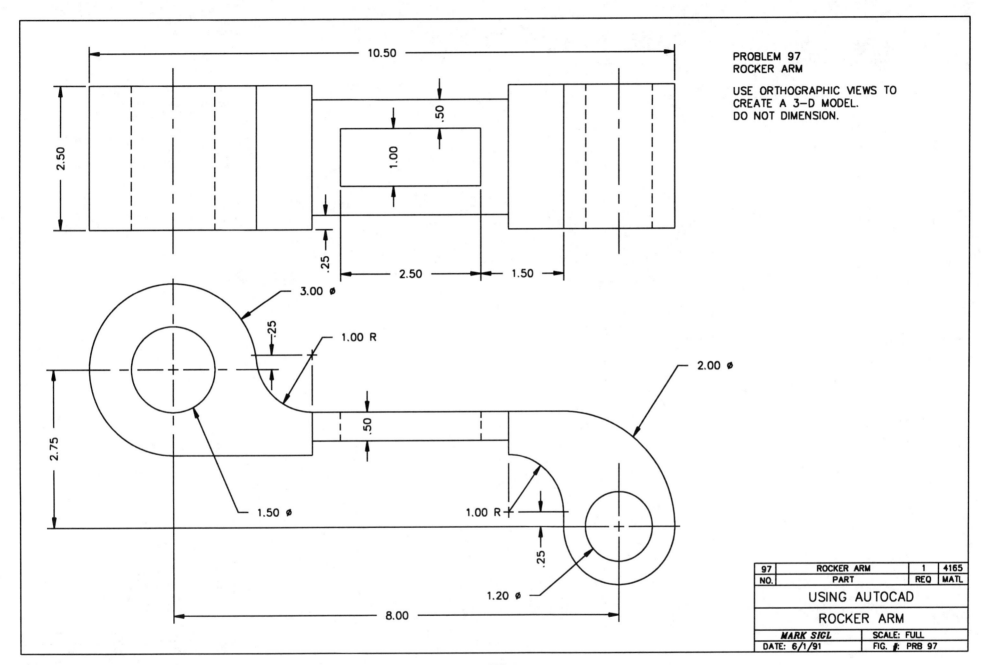

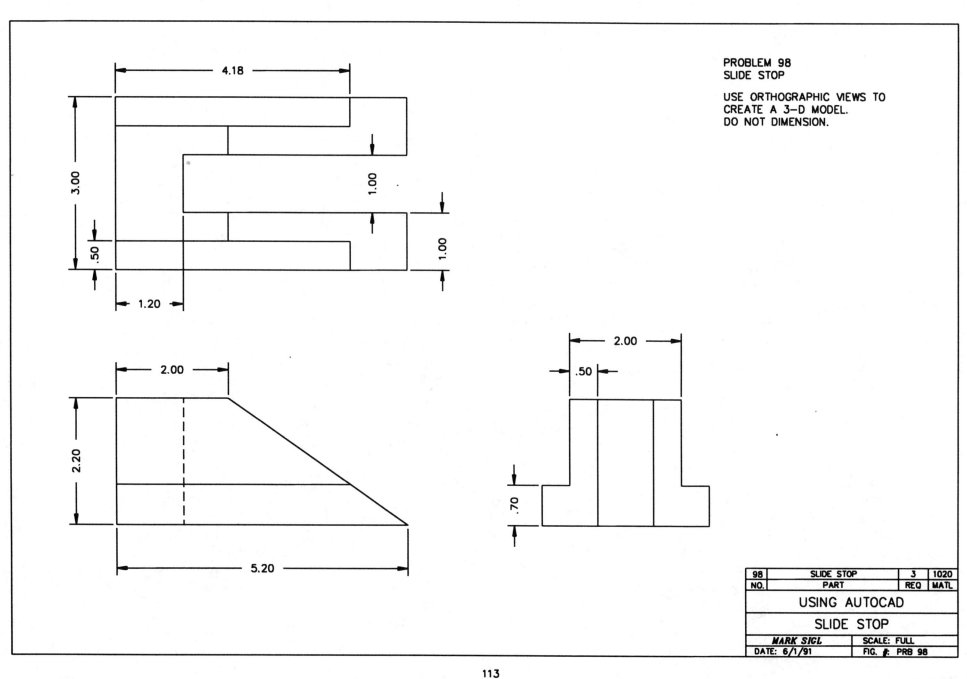

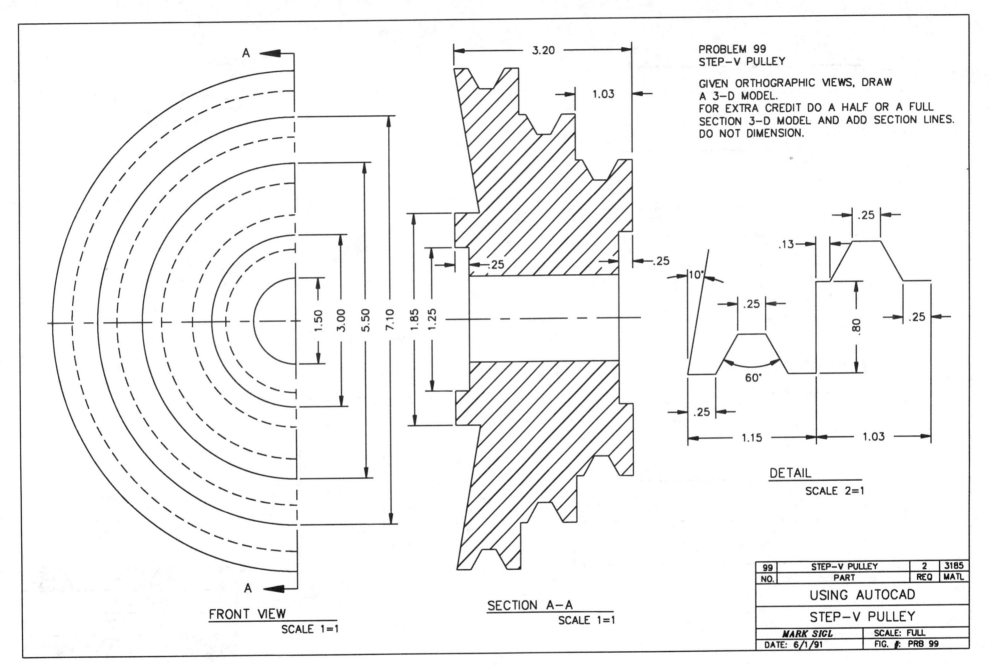

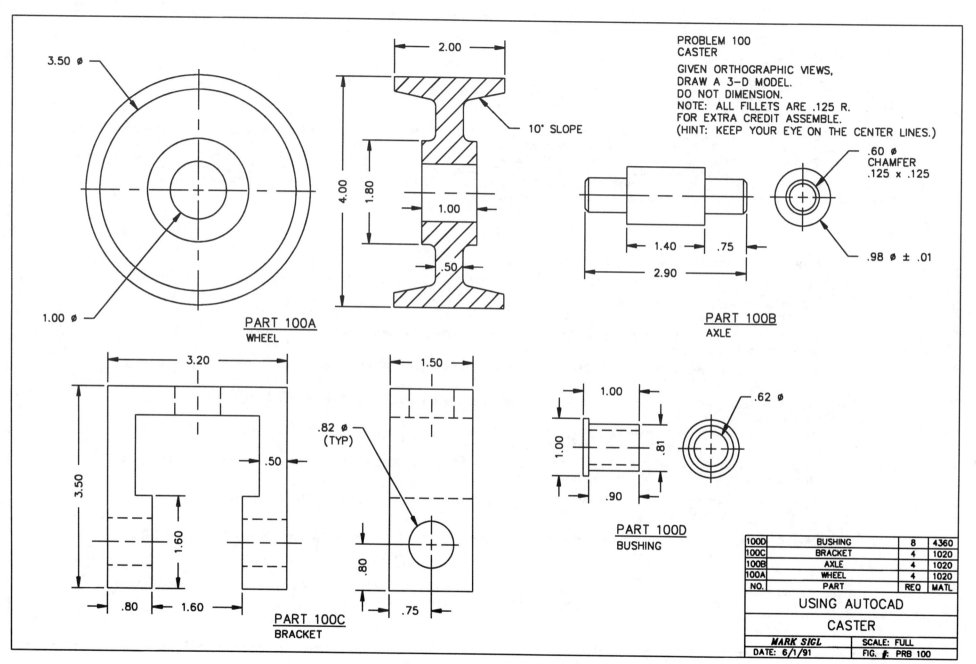

# CHAPTER 11
# SOLID MODELING
# ADVANCED MODELING EXTENSION (A.M.E.)

THIS CHAPTER EXPLORES THE USE OF SOLID 3-DIMENSIONAL MODELING. SOLID 3-DIMENSIONAL MODELING CAN BE UTILIZED TO PERFORM MANY FUNCTIONS SUCH AS: STRESS ANALYSIS, HEAT TRANSFER, COMPUTER-AIDED MANUFACTURING, AND GRAPHIC RENDERINGS. COMPLEX SOLID MODELS CAN BE COMBINING OR SUBTRACTING SIMPLE SOLID OBJECTS. BE SURE TO USE AS MANY ADVANCED MODELING EXTENSION COMMANDS AS POSSIBLE INCLUDING: SOLBOX, SOLWEDGE, SOLCONE, SOLCYL, SOLSPHERE, SOLTORUS, SOLFILL, SOLCHAM, SOLEXT, SOLREV, SOLIDIFY, SOLUNION, SOLINT, SOLSUB, SOLSEP. ALSO MAKE USE OF COMMANDS RELATED TO MODIFICATION, QUERY, DOCUMENTATION AND REPRESENTATION. SOLLIST, SOLMASSP, SOLAREA, SOLVAR, SOLCHP, SOLMAT, SOLMOVE, SOLFEAT, SOLPROF, SOLUCS, SOLWIRE, SOLMESH.

PROBLEM                                             TOPIC

101         GUIDE PLATE EXPOSES YOU TO THE BASIC SOLID COMMANDS.

102         THE TEE STOPS CONTINUES THE DEVELOPMENT OF SOLID MODELING SKILLS.  FILLETS TO PERPENDICULAR EDGES PROVIDE ADDITIONAL CHALLENGES.

103         IRREGULAR SURFACES AND EDGES ARE COVERED WITH THE SHAFT BLOCK GUIDE. USE SOLMAT, SOLAREA, AND SOLMASSP TO FIND
            THE MASS, VOLUME AND AREA OF THE OBJECT DRAWN.

104         COLLAR GUIDE CHALLENGES YOU WITH SLANTED SURFACES AND CIRCLES AND ALSO GIVES YOU EXPERIENCE IN
            SECTIONING A 3-D MODEL.

105         OBLIQUE HOLES ARE INCLUDED IN THE GUIDE BLOCK.  THIS COMPLEX OBJECT SHOULD STRENGTHEN YOUR'S 3-D CONSTRUCTION
            TECHNIQUES.

106         THE SHAFT GUIDE INCORPORATES 3-D FILLETS AND TANGENCIES.  THIS UNIQUE OBJECT HELPS YOU GAIN EXPERIENCE
            CONSTRUCTING 3-D PYRAMIDS.

107         THE LINK UNIT WILL DEVELOP YOUR CONSTRUCTION SKILLS IN BUILDING MODELS CONTAINING CURVES AND ARCS.
            REMEMBER TO APPLY SUBTRACTION OF SOLID SURFACES.

108         THE HOUSING UNIT IS COMPOSED OF TWO PARTS TO BE DRAWN AS A SOLID MODEL.  IN ADDITION, SHOW THE UNIT ASSEMBLED
            AND COMPLETE FULL SECTION OF THE ASSEMBLED UNIT.

109         HOUSING BLOCK OFFERS MANY UNIQUE CHALLENGES.  THIS SHOULD NOT BE A PROBLEM AFTER COMPLETING
            THE EARLY 3-D WIRE FRAME MODEL PROBLEMS.

110         FINAL CHALLENGE.  DRAW THE HAND CLAMP IN 3-D AND OBTAIN THE ORTHOGRAPHIC VIEWS. DIMENSION, AND ASSEMBLE THE DRAWING..

NOTE: FOR ADDITIONAL CHALLENGES YOU CAN RETURN TO PREVIOUS CHAPTERS AND CREATE 3-DIMENSIONAL MODELS OF THE DRAWINGS.
GOOD LUCK WITH YOUR FUTURE CADDING EXPERIENCES!

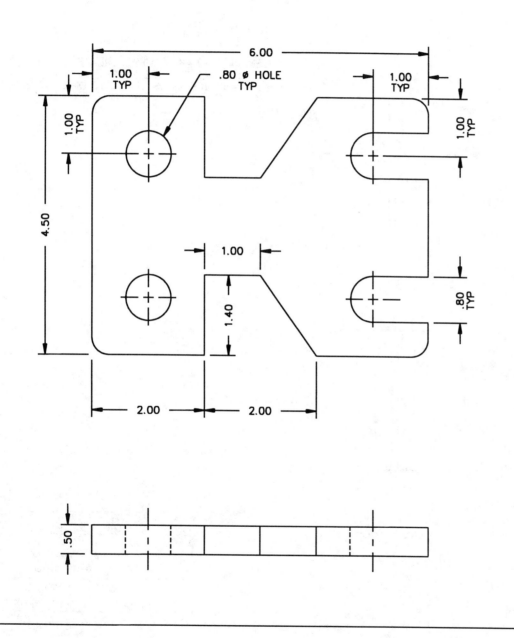

PROBLEM 101
GUIDE PLATE

GIVEN ORTHOGRAPHIC VIEWS CREATE
A 3-D MODEL OF THE PLATE.
ADD A 1/16 CHAMFER TO BOTH SIDES, OBTAIN
FRONT AND TOP VIEWS WITH DIMENSIONS.
ALSO ADD AN ISOMETRIC VIEW.
NOTE: FILLETS ARE .30 R.

| 101 | GUIDE BLOCK | 1 | AL |
|---|---|---|---|
| NO. | PART | REQ | MATL |

USING AUTOCAD

GUIDE PLATE

| MARK SIGL | SCALE: FULL |
|---|---|
| DATE: 7/1/92 | FIG. #: PRB 101 |

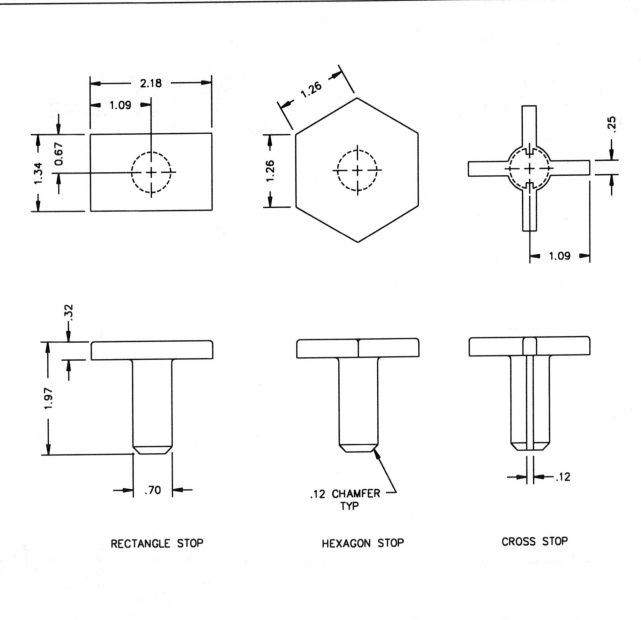

PROBLEM 102
TEE STOPS

GIVEN ORTHOGRAPHIC VIEWS CREATE A
3-D MODEL OF THE TEE STOPS. OBTAIN
FRONT AND TOP VIEWS INCLUDING DIMENSIONS.
ALSO INCLUDE A PICTORIAL VIEW OF EACH STOP.
NOTE: FILLETS ARE .06 R.

| NO. | PART | REQ | MATL |
|---|---|---|---|
| 102A | TEE STOP – HEXAGON | 100 | STEEL |
| 102B | TEE STOP – HEXAGON | 30 | STEEL |
| 102C | TEE STOP – CROSS | 59 | STEEL |

USING AUTOCAD
TEE STOPS
MARK SIGL   SCALE: FULL
DATE: 7/1/92   FIG. #: PRB 102

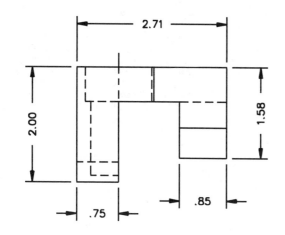
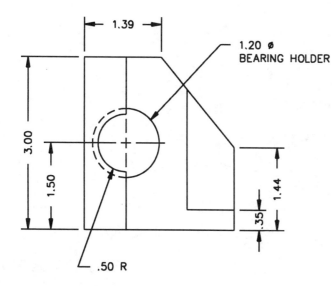
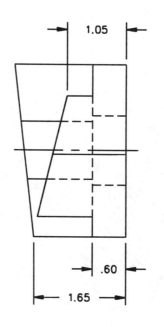

PROBLEM 103
SHAFT BLOCK GUIDE

GIVEN ORTHOGRAPHIC VIEWS CREATE
A 3-D MODEL OF SHAFT BLOCK GUIDE.
OBTAIN THREE VIEWS WITH DIMENSIONS AND
PLACE A 3-D VIEW IN UPPER RIGHT
RIGHT HAND CORNER.
USE SOLMAT, SOLAREA, AND SOLMASSP TO
OBTAIN FOLLOWING INFORMATION:

MASS:
VOLUME:
BOUNDING BOX:
AREA:

| 103 | SHAFT BLOCK GUIDE | 4 | AL |
|---|---|---|---|
| NO. | PART | REQ | MATL |

USING AUTOCAD

SHAFT BLOCK GUIDE

| *MARK SIGL* | SCALE: FULL |
|---|---|
| DATE: 7/1/92 | FIG. #: PRB 103 |

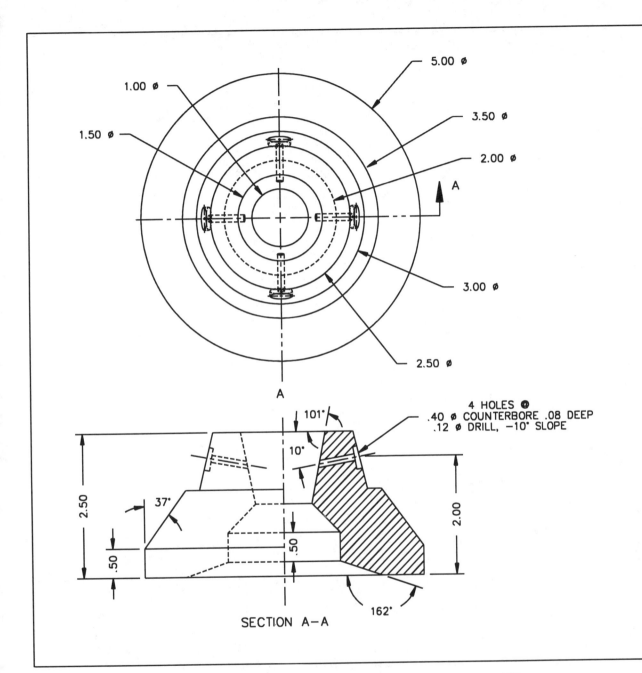

PROBLEM 104
COLLAR GUIDE

GIVEN ORTHOGRAPHIC VIEWS AND
HALF SECTION, CREATE A 3-D MODEL
OF COLLAR GUIDE. OBTAIN FRONT
AND TOP VIEWS WITH DIMENSIONS.
ALSO COMPLETE AN ISOMETRIC VIEW WITH A
HALF SECTION.
NOTE: ALL SURFACES ARE MACHINED.

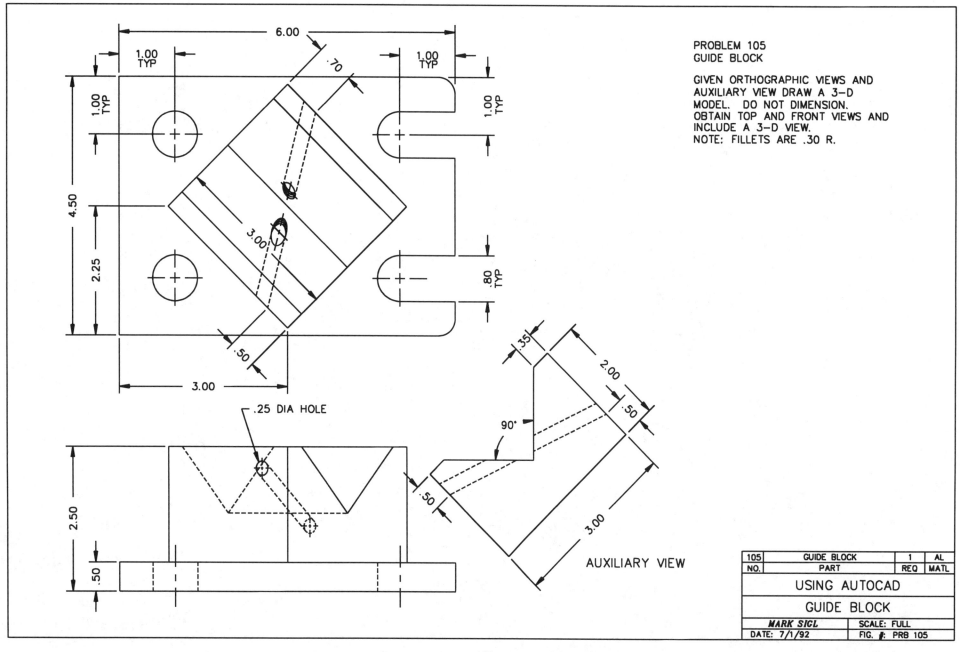

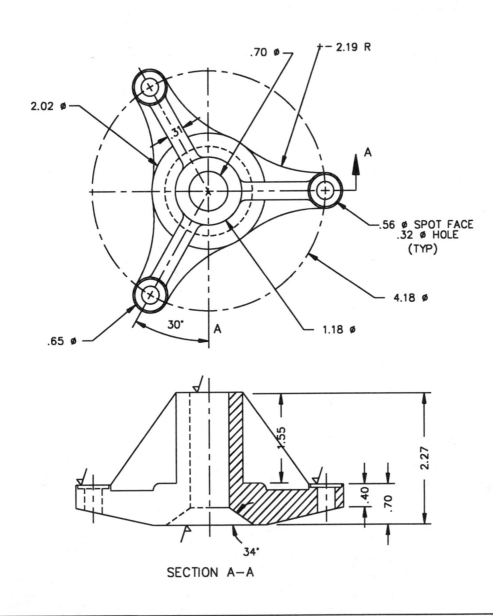

PROBLEM 106
SHAFT GUIDE

GIVEN ORTHOGRAPHIC VIEWS AND
SECTION VIEW OF SHAFT GUIDE,
CREATE A 3-D MODEL OF GUIDE.
DO NOT FORGET ABOUT FILLETS.
OBTAIN FRONT AND TOP VIEWS WITH DIMENSIONS
AND PLACE A HALF-SECTIONED ISOMETRIC VIEW.

NOTE:

1. FILLETS ARE .125 R.
2. √ DENOTES FINISHED SURFACE MACHINED .032."

| 106 | SHAFT GUIDE | 4 | CI |
|---|---|---|---|
| NO. | PART | REQ | MATL |

USING AUTOCAD

SHAFT GUIDE

| MARK SIGL | SCALE: FULL |
|---|---|
| DATE: 7/1/92 | FIG. #: PRB 106 |

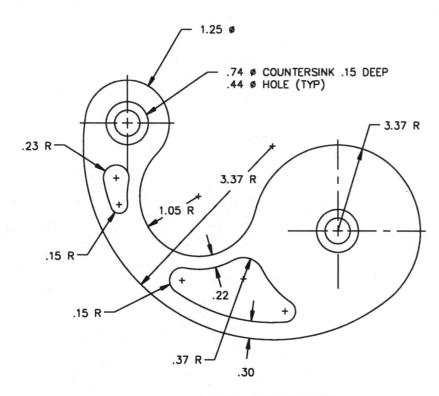

PLATE — RIGHT SIDE
PLATE THICKNESS IS .30"

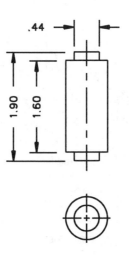

SPACING POSTS

PROBLEM 107
LINK UNIT

GIVEN ORTHOGRAPHIC VIEWS CREATE
A 3-D MODEL OF LINK UNIT.
OBTAIN FRONT AND TOP VIEWS WITH
DIMENSIONS AND PLACE A 3-D VIEW.

NOTE:

1. ALL HOLES ARE COUNTERSUNK .15".
2. ALL SURFACES ARE MACHINED .030 in.

| 107A | LINK UNIT | 2 | SS |
| 107B | SPACING POSTS | 2 | SS |
| NO. | PART | REQ | MATL |

USING AUTOCAD
LINK UNIT

MARK SIGL | SCALE: FULL
DATE: 7/1/92 | FIG. #: PRB 107

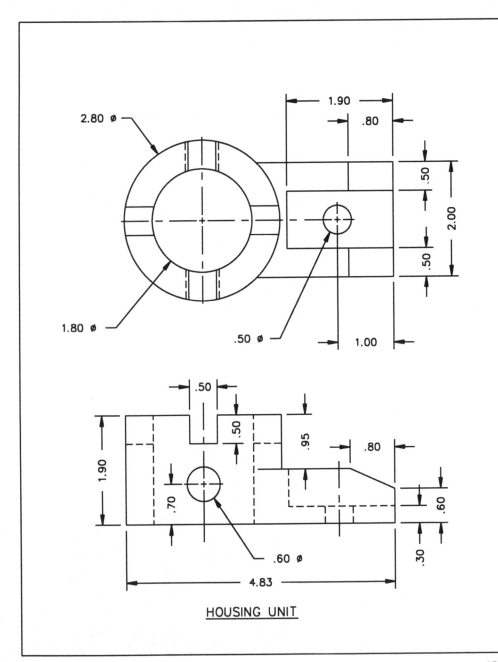

HOUSING UNIT

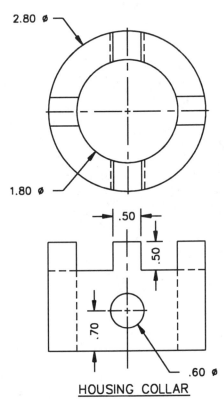

HOUSING COLLAR

PROBLEM 108
HOUSING UNIT

GIVEN ORTHOGRAPHIC VIEWS CREATE
A 3-D MODEL OF HOUSING UNIT.
OBTAIN FRONT AND TOP VIEW WITH DIMENSIONS
AND AN ASSEMBLED 3-D VIEW AND A
A FULL SECTION OF ASSEMBLED 3-D VIEW.
NOTE: ALL SURFACES ARE MACHINED .030 in.

| 108A | HOUSING UNIT | 1 | SS |
| 108B | HOUSING COLLAR | 1 | SS |
| NO. | PART | REQ | MATL |

USING AUTOCAD

HOUSING UNIT

MARK SIGL — SCALE: FULL
DATE: 7/1/92 — FIG. #: PRB 108

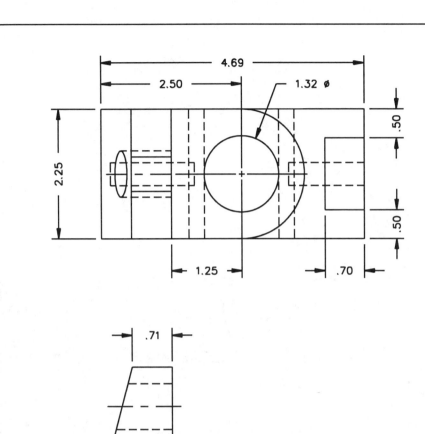

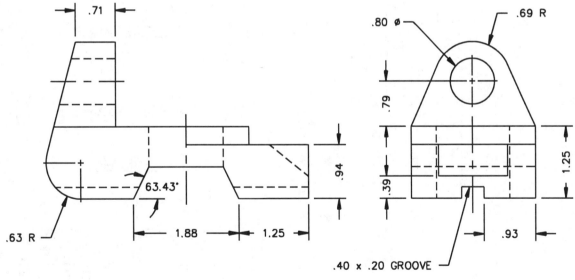

HOUSING BLOCK

PROBLEM 109
HOUSING BLOCK

GIVEN ORTHOGRAPHIC VIEWS CREATE
A 3-D MODEL OF HOUSING BLOCK.
OBTAIN FRONT, TOP, AND RIGHT SIDE VIEWS
WITH DIMENSIONS. ALSO INCLUDE A FULL
SECTION OF A 3-D VIEW.
NOTE: ALL SURFACES ARE MACHINED .030 in.

| 109 | HOUSING BLOCK | 1 | AL |
|---|---|---|---|
| NO. | PART | REQ | MATL |

USING AUTOCAD

HOUSING BLOCK

| MARK SIGL | SCALE: FULL |
|---|---|
| DATE: 7/1/92 | FIG. #: PRB 109 |

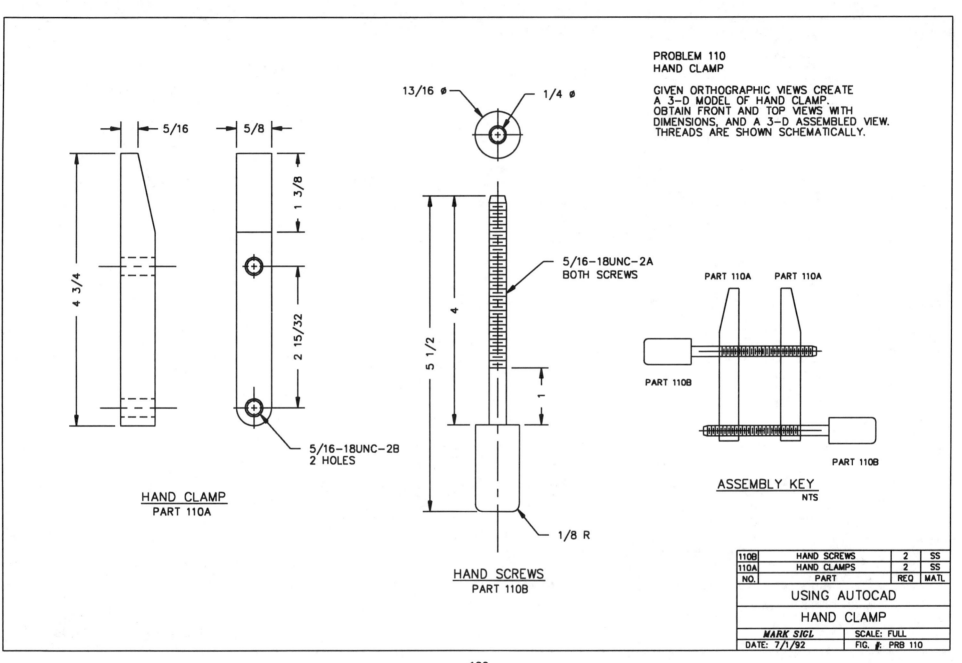

PROBLEM A-1
DRAWING TEST

INSTRUCTIONS: DRAW THE PLATE SHOWN BELOW.  DO NOT DIMENSION.
ALSO, GIVE EACH CORNER A RADIUS OF .35 R.
BE SURE TO SAVE YOUR DRAWING. THE TIME IS RECORDED IN
THE DRAWING FILE. (USE THE TIME COMMAND TO LOOK AT.)

TIME LIMIT: 20 MINUTES

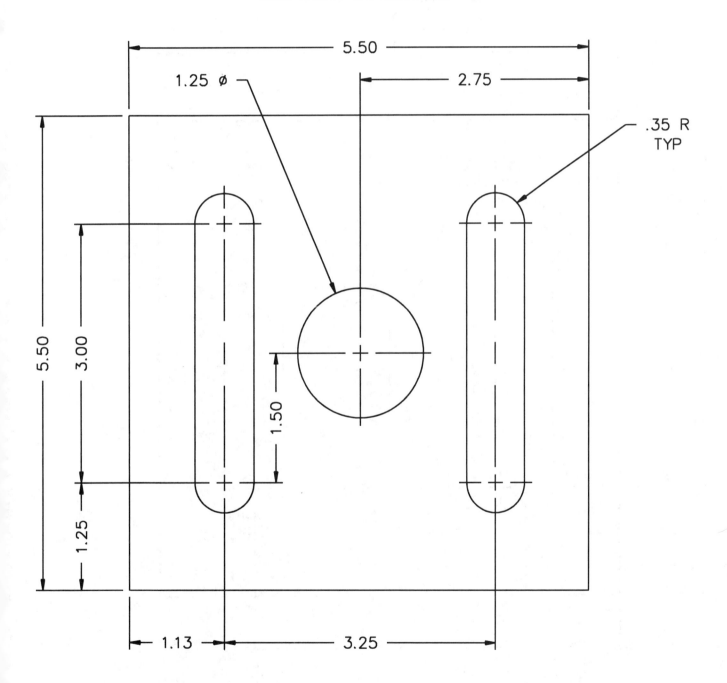

PROBLEM A-2
EDITING TEST

INSTRUCTIONS: MANIPULATE ONE-INCH SQUARE AND ONE CIRCLE USING YOUR EDIT COMMANDS ONLY TO CREATE THE PLATE SHOWN BELOW. DO NOT DIMENSION. ALSO, GIVE EACH CORNER A .35 RADIUS. REMEMBER TO SAVE YOUR DRAWING SO YOU CANNOT STRETCH THE FACTS.

TIME LIMIT: 30 MINUTES

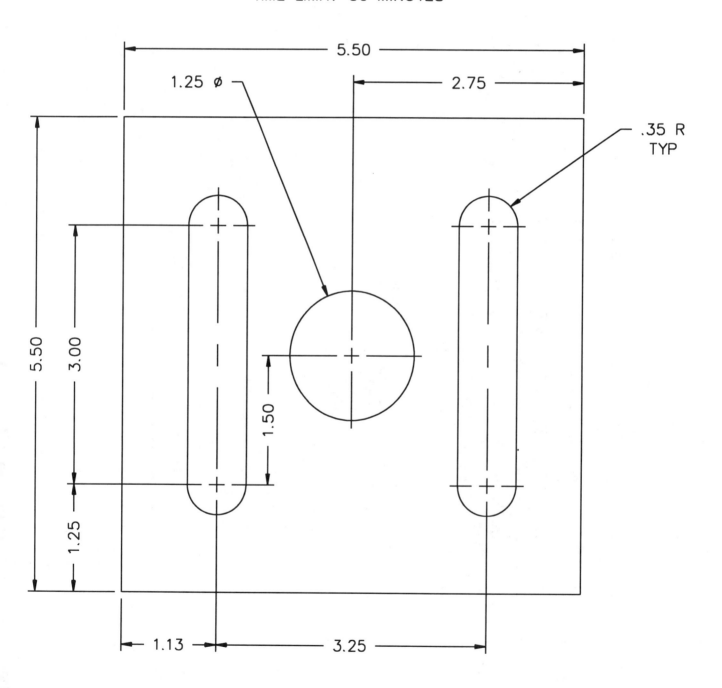

PROBLEM A-3
ARRAYING TEST

INSTRUCTIONS: DRAW THE SPROCKET AND DO NOT DIMENSION.
BE SURE TO SAVE YOUR DRAWING. THE TIME IS RECORDED IN
THE DRAWING FILE. (USE THE TIME COMMAND TO LOOK AT.)

TIME LIMIT: 20 MINUTES

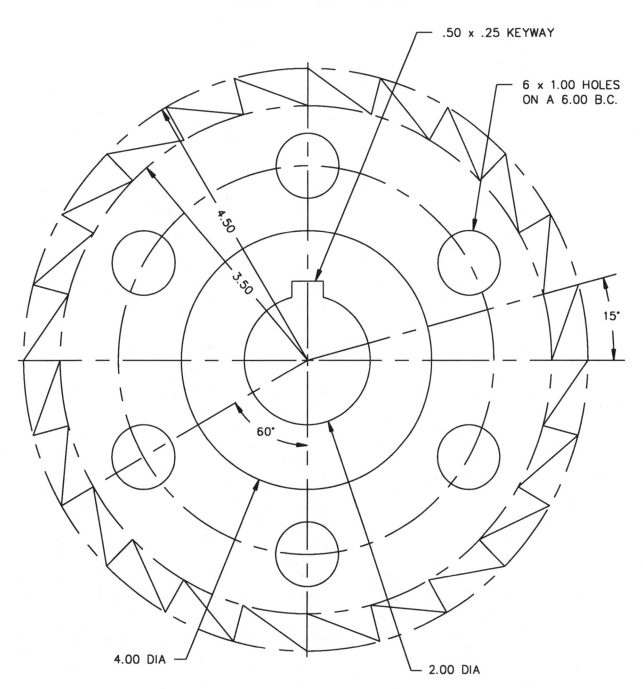

PROBLEM A-4
DIMENSIONING TEST

INSTRUCTIONS: RECALL PROBLEM 14 AND PROVIDE THE COMPLETE DIMENSIONS.

TIME LIMIT: 20 MINUTES

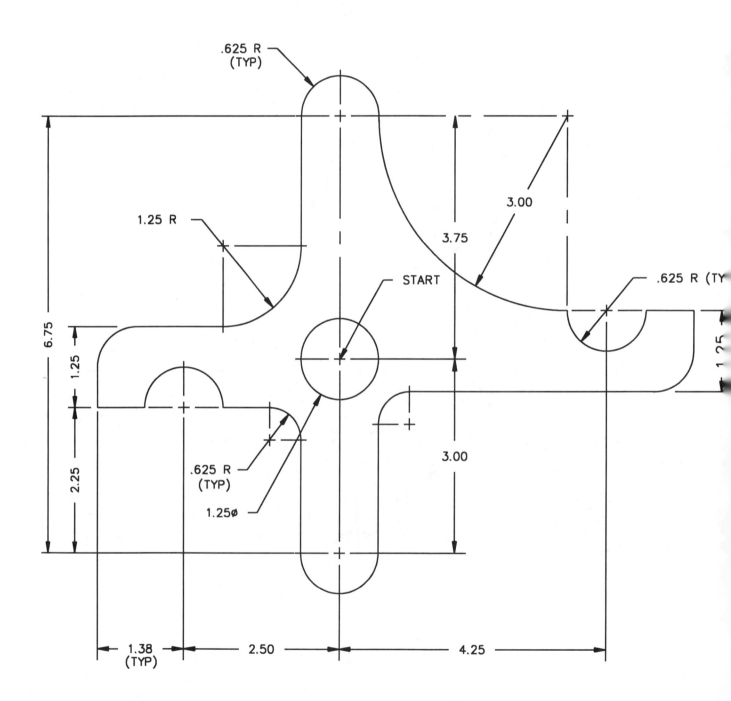

PROBLEM A-5
DIGITIZING TEST

DRAW A SIX-INCH SQUARE TO BE USED AS A REFERENCE MEASURE
AND DIGITIZE THE FOLLOWING OBJECTS TO CHECK FOR ACCURACY.
LIST THE INFORMATION TO NEAREST THOUSANDTHS.
GOOD LUCK IN DIGITIZING AND BEING ACCURATE, IT IS A SIGN
OF A GOOD CAD OPERATOR.

TIME LIMIT: 30 MINUTES

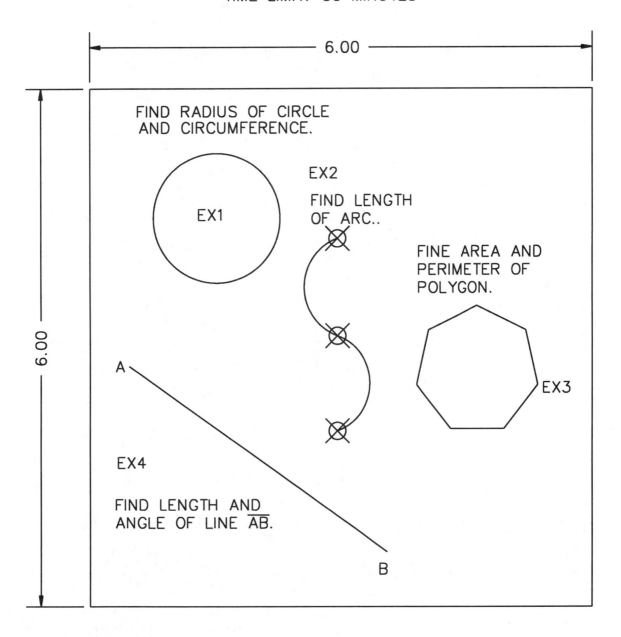

PROBLEM A-6
INQUIRY DRAWING TEST

DRAW THE GASKET BELOW INCLUDING DIMENSIONS. FIND GROSS, HOLE, AND NET AREA OF THE GASKET TO NEAREST THOUSANDTHS.

TIME LIMIT: 30 MINUTES

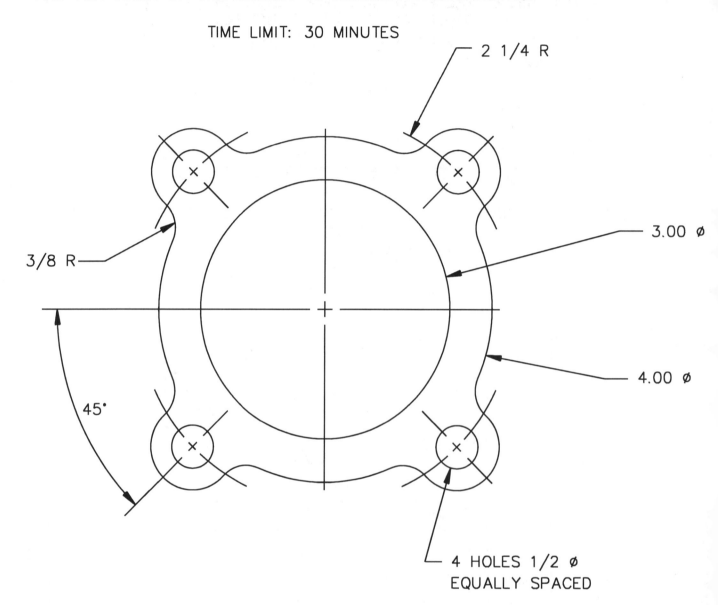

## AREA CALCULATIONS

| GROSS AREA: | 15.3227 |
|---|---|
| HOLE AREA: | − 7.8540 |
| NET AREA: | 7.4687 |

## NOTES:

1. GASKET MATERIAL TO BE .030 THICK.
2. ANGLES TO BE ±1/2°.
3. ALL DIMENSIONS TO BE ±.002 (UNLESS OTHERWISE SPECIFIED).

# REVIEW QUESTIONS A

Score:_____                                    Name:_____

MULTIPLE CHOICE.        Select the best answer.

1.   Which single key allows you to turn ORTHO mode ON/OFF?

   A.   F6
   B.   F7
   C.   F8
   D.   F9


2.   How can an AutoCAD command be repeated from the keyboard without reentering them?

   A.   Enter key
   B.   Space bar
   C.   ESC key
   D.   A and B


3.   What command saves your drawing to a filename without returning you to the Main Menu
     and allows you the option of renaming the current filename?

   A.   Quit
   B.   End
   C.   Save
   D.   Exit


4.   What happens when you pick AutoCAD from the screen menu?

   A.   Takes you to the Root Page of the menu
   B.   Enters Root Page after the command prompt
   C.   Exits you to the Main Menu
   D.   None of the above


5.   What happens when you use the "U" option while drawing a line segment?

   A.   Deletes all line segments
   B.   Exits the Line command without altering segments drawn
   C.   Has no effect on line segments
   D.   Backs up or undoes last Line segment drawn


6.   What happens when you use the Arc Continue option?

   A.   Continues an Arc immediately after drawing a Line
   B.   Continues an Arc Tangent to the previous arc
   C.   Increases the Radius of an Arc
   D.   A and B


7.   What direction does a positive number indicate when specifying angles in degrees?

   A.   Clockwise
   B.   Counterclockwise
   C.   Has No impact when specifying angles in degrees
   D.   None of the above

8.  What kind of Entities are effected by the Drag Mode?

    A.  Drawing Circles and Arcs
    B.  Drawing Lines and Polygons
    C.  Drawing Ellipses and Donuts
    D.  All of the above


9.  Regarding Arc options, what does "S.C.E." mean?

    A.  Start, Center, End
    B.  Second, Continue, Extents
    C.  Second, Center, End
    D.  Start, Continue, End


10. What command is used to Remove Entities from the drawing?

    A.  Del
    B.  Erase
    C.  U
    D.  All of the above


11. Will OOPS work if you draw something **AFTER** erasing the object?

    A.  No
    B.  Yes
    C.  Sometimes
    D.  Only the CAD-Pro can make it work


12. What command is used to draw circle segments?

    A.  Segment
    B.  Circle
    C.  Ellipse
    D.  Arc


13. The command @5.31<45 is an example of

    A.  Absolute coordinate
    B.  Relative coordinate
    C.  Polar coordinate
    D.  Angular coordinate


14. Which command forces "Picked" Points to automatically align with an Imaginary Grid?

    A.  Grid
    B.  Snap
    C.  Osnap
    D.  Axis


15. The _____ command places a Ruler Line on the edge of the screen.

    A.  Grid
    B.  Snap
    C.  Ortho
    D.  Axis

16. The _____ command forces all lines to be drawn only horizontally or vertically.

    A. Grid
    B. Snap
    C. Ortho
    D. Osnap

17. The _____ command reverses several command moves at once.

    A. U
    B. Undo
    C. Redo
    D. Oops

18. The _____ command is used to "xerox" entities and place them in another location.

    A. Offset
    B. Move
    C. Copy
    D. Mirror

19. The _____ allows the creation of an entity parallel to the original.

    A. Offset
    B. Move
    C. Copy
    D. Mirror

20. The _____ command allows you to store a screen display identified by a name.

    A. Block
    B. View
    C. Zoom P
    D. ID

21. Using the Scale command, what number would you enter to enlarge an object by 50% ?

    A. 0.5
    B. 3
    C. 1.5
    D. 50

22. The _____ command is a mode that causes all text to be replaced by a box that approximates the shape of the text string.

    A. Style
    B. QText
    C. Box
    D. DText

23. Which statement best describes the Divide command?

    A. Displays absolute coordinates and current layer of chosen entity
    B. Calculates total length of an entity
    C. Separates an entity into an equal number of parts
    D. Divides an entity into intervals of specified lengths

135

24. Which commands would you be most likely to use to remove a line segment drawn too far past an intersecting line?

    A. Cut
    B. Break
    C. Trim
    D. Remove

25. With regard to the Solid command which of the following is (are) True?

    A. Points must be selected in a certain order.
    B. Order of selection is unimportant.
    C. Fill must be ON in order to use SOLID.
    D. Both A and C.

MATCHING. Match Zoom command options with their descriptions.

| | |
|---|---|
| ___26. Option based on previous screen center point; specifying a number only calculates amount of reduction or enlargement of display according to its initial size. Adding an X after the number calculates display magnification from current display. | A.    ALL<br><br>B.    WINDOWS<br><br>C.    EXTENTS<br><br>D.    SCALE |
| ___27. Causes entire drawing to be displayed on screen. This shows the limits or if the drawing size exceeds the limits, it also shows the entities of the drawing lying outside the limits. | E.    LEFT CORNER<br><br>F.    MAGNIFICATION |
| ___28. Shows all drawing entities at their maximum screen size. | G.    DYNAMIC |
| ___29. Shows that portion of the drawing "windowed" at its maximum screen size. | H.    CENTER |

___30. Allows selection of a new display center point and a new magnification value; new display is centered on new center point.

___31. Magnifies display based on lower left corner; X option is valid.

___32. Provides informational display showing current generated area, drawing extents, view box, and current view window.

___33. Asks for a numeric magnification factor.

MATCHING.    Match special characters with their TEXT code.

| | | | |
|---|---|---|---|
| ___34. overscore | | A. | %%d |
| ___35. underscore | | B. | %%c |
| ___36. degrees symbol | | C. | %%p |
| ___37. plus/minus tolerance symbol | | D. | %%o |
| ___38. circle diameter dimensioning symbol | | E. | %%u |

MATCHING.     Match distance measures available in UNITS with their format.

___39.  Scientific                                              A.      1'–9.75"

___40.  Decimal                                                 B.      21 3/4"

___41.  Engineering                                             C.      21.75

___42.  Architectural                                           D.      2.175E+1

___43.  Fractional                                              E.      1'–9 3/4"


MATCHING.     Match LAYER option with its correct description.

___44.  Provides a listing of the defined layer           A.      ON
        including layer name, state, color, and linetype.
                                                          B.      SET
___45.  Combines actions of the NEW and SET options.
                                                          C.      THAW
___46.  Makes named layer current.
                                                          D.      MAKE
___47.  Names the layer.
                                                          E.      FREEZE
___48.  Makes named layer visible.
                                                          F.      OFF
___49.  Makes the named layer invisible.
                                                          G.      LTYPE
___50.  Defines the color for a chosen layer.
                                                          H.      ?
___51.  Defines a linetype for the layer.
                                                          I.      NEW
___52.  Makes the layer invisible and prevents
        regeneration of the named layer.                  J.      COLOR

___53.  Makes the layer visible and enables
        regeneration of the named layer.


MULTIPLE CHOICE.     Select the best answer.


54.  The _____ command removes entities from the drawing.

     A.  OOPS
     B.  ERASE
     C.  DELETE
     D.  LAST


55.  What command undoes the last erase that was made?

     A.  OOPS
     B.  LAST
     C.  UNERASE
     D.  DARN!

56. The line command @ 6.75<60 is an example of a:

   A. Absolute coordinate
   B. Relative coordinate
   C. Polar coordinate
   D. Angular coordinate

57. The _____ command forces "picked" points to automatically align with an imaginary grid.

   A. GRID
   B. SNAP
   C. OSNAP
   D. AXIS

58. The _____ command places a ruler line on the edge of the screen.

   A. GRID
   B. SNAP
   C. ORTHO
   D. AXIS

59. The _____ command reverses several command moves at once.

   A. U
   B. UNDO
   C. REDO
   D. OOPS

60. The _____ command is used to erase part of a line, arc, circle, or trace.

   A. ERASE
   B. BREAK
   C. TRIM
   D. EXTEND

61. The _____ command is used to modify existing entities.

   A. CHANGE
   B. MOVE
   C. COPY
   D. MIRROR

62. The _____ command allows you to change location of entities.

   A. CHANGE
   B. MOVE
   C. COPY
   D. MIRROR

63. The _____ allows creation of an entity parallel to the original.

   A. OFFSET
   B. MOVE
   C. COPY
   D. MIRROR

64. The _____ command is used to make multiple copies of one or more objects in rectangular or polar patterns.

    A. ARRAY
    B. MOVE
    C. COPY
    D. MINSERT

65. The _____ command allows you to change the scale of a(n) entity(ies). The X and Y scales are changed equally.

    A. ROTATE
    B. SCALE
    C. TRIM
    D. EXTEND

66. The _____ command allows you to rotate an entity or group of entities around a chosen base point.

    A. ROTATE
    B. SCALE
    C. TRIM
    D. EXTEND

67. The _____ command allows you to trim objects in a drawing by defining other objects as cutting edges, then specifying the part of the object to be cut.

    A. STRETCH
    B. SCALE
    C. TRIM
    D. EXTEND

68. The _____ command allows you to extend objects in a drawing to meet a boundary entity.

    A. ROTATE
    B. SCALE
    C. TRIM
    D. EXTEND

69. The _____ command causes AutoCAD to recalculate and redisplay the drawing database.

    A. REDRAW
    B. REGEN
    C. RECALC
    D. SCALE

70. The _____ command allows you to temporarily exit AutoCAD in order to use all DOS commands.

    A. EXIT
    B. SH
    C. SHELL
    D. END

71. The _____ command allows you to store a screen display identified by a name.

    A. BLOCK
    B. VIEW
    C. ZOOM P
    D. ID


72. A _____ is a pattern from which AutoCAD constructs text.

    A. STYLE
    B. QTEXT
    C. FONT
    D. DTEXT


73. A _____ is a collection of modifiers applied to the font of your choice.

    A. STYLE
    B. QTEXT
    C. FONT
    D. DTEXT


74. What command is used to create a new text style?

    A. STYLE
    B. QTEXT
    C. FONT
    D. DTEXT


75. If you enter 72%%% on the text line while in TEXT, what would be generated on the screen?

    A. 72%%%
    B. 72<DIAMETER SIGN>
    C. 72<DEGREE SIGN>
    D. 72%


76. The _____ command allows the CAD user to set the drawing units appropriate for the type of drawing.

    A. LIMITS
    B. SCALE
    C. UNITS
    D. TYPE


77. The basic definition of a layer **DOES NOT** include which of the following?

    A. Layer name
    B. Layer color
    C. Layer linetype
    D. Layer number


78. Which best describes the STATUS command?

    A. Overrides layer definition of a line type.
    B. Causes new entities to be drawn in color defined by the layer definition.
    C. Rests scale that defines how a line type is drawn.
    D. Allows you to view current drawing modes, defaults,
       limits, and other parameters in text form.

140

79. Which statement best describes the diameter option of the DIM command?

    A. Draws an arc to show angle between two lines.
    B. Draws a dimension line across the diameter of an arc/circle.
    C. Draws a dimension line like a radius in an arc or circle.
    D. Draws only one dimension and then returns you to the command prompt.


80. Which of the following statements about LIMITS are **TRUE**?

    A. Attempts to draw outside the preset boundaries will be rejected.
    B. Limits determine the area in which the grid will be shown.
    C. The limits determine the area ZOOM Extents will display.
    D. A and B.


81. Which UNITS option is shown improperly?

    A. Scientific: 2.175E+1
    B. Decimal: 1'-9.75"
    C. Architectural: 1'-9 3/4"
    D. Fractional: 21 3/4"


82. Which LAYER option is matched with an incorrect description?

    A. SET: Makes the named layer current.
    B. NEW: Names the layer.
    C. ON: Makes the named layer visible.
    D. OFF: Makes the layer invisible and prevents regeneration of the named layer.


83. Which statement best describes the LIST command?

    A. Displays the absolute coordinates and current elevation of the chosen point.
    B. Calculates the perimeter, in the set drawing units, for a closed polygon defined by user generated points.
    C. Used to display the data stored by AutoCAD about any selected entity.
    D. Causes all the data stored by AutoCAD about each drawing entity to be listed.


84. Which statement best describes the DIVIDE command?

    A. Displays the absolute coordinates and current elevation of the chosen point.
    B. Calculates the perimeter, in the set drawing units, for a closed polygon defined by user generated points.
    C. Separates an entity into an equal number of parts.
    D. Divides an entity into intervals of specified lengths.


85. The BLOCK command:

    A. Is a group of entities identified by a name.
    B. Is used to identify a group of entities and to name them.
    C. Takes the entities defined as a block and writes them to the disk as a separate drawing.
    D. Is used to define an insertion base point for a drawing file other than the default value (0,0).


86. The HATCH command:

    A. Is a pattern of lines, dots, and dashes that make up a defined pattern.
    B. Is used to fill a specified polygon with a picked pattern using the chosen style.
    C. Allows freehand drawing in AutoCAD.
    D. Is used to fill a polygon with a solid color.

87. To enter the Isometric Mode, issue the _____ command.

   A. GRID
   B. ELLIPSE
   C. ISOPLANE
   D. SNAP


88. To change from one isometric axis to another, use:

   A. CTRL E
   B. ISOPLANE command.
   C. GRID
   D. A and B.


89. Which of the following are **FALSE**?

   A. At the "Select Polyline" prompt in PEDIT, the line selected does not have to be a polyline.

   B. Vertex editing is used to modify the vertices and the segments between chosen vertices of a polyline.

   C. The DIVIDE command works only on lines, arcs, circles, or polylines.

   D. A block is not stored as part of the drawing file.


90. Which of the following are **FALSE**?

   A. The block insertion base point is the point chosen by the operator to define where the block is placed in reference to the block insertion point.

   B. Entities to be defined as a block can be chosen by pointing.

   C. When the BLOCK command is completed, the entities selected are erased from the screen.

   D. A block created on the 0 layer will, when inserted, be placed on the current layer.


91. Which of the following are **FALSE**?

   A. By issuing the BLOCK command and entering "?" to the prompt, all blocks associated with the drawing are listed.

   B. The INSERT command can be used to placed a block, a file created by WBLOCK, or a drawing file in the current drawing.

   C. During block insertion, placing an asterisk (*) in front of the block name to be inserted has no effect.

   D. The EXPLODE command breaks a block into its component parts.


92. Which PLINE option and its description is incorrect?

   A. Close: Works like the CLOSE command in LINE.
   B. Dist: Draws a line segment at the same angle as previous segment for the specified length.
   C. Undo: Last polyline entered is undone.
   D. Width: Selects width of polyline.

142

93. Which PEDIT option and its description is incorrect?

   A. Open: Converts and connects nonpolyline entities to polylines.
   B. Width: Allows selection of new width for all segments of polyline.
   C. Fit Curve: Constructs a smooth curve using the vertices.
   D. X: Returns to command prompt.


94. The ERASE command:

   A. Erases part of a entity.
   B. Erases any entity.
   C. Erases any group of entities.
   D. Both B and C.


95. The OFFSET command:

   A. Produces a mirror image of all objects.
   B. Contains a solid fill option.
   C. Allows the creation of an entity parallel to another entity.
   D. None of the above.


96. AutoCAD text can be:

   A. Left and right justified.
   B. Centered and aligned between two points.
   C. Created as large or as small as desired.
   D. All of the above.


97. Which one of the following commands is used to erase a small piece of a polyline?

   A. PEDIT
   B. ERASE
   C. UNDO
   D. BREAK


98. The common element of both END and QUIT commands is that:

   A. both save your work.
   B. neither saves your work.
   C. both exits the Drawing Editor.
   D. neither is useful.


99. What command is used to turn the dialogue box on and off?

   A. ATTDISP
   B. ATTDIA
   C. ATTDEF
   D. DIALOGUE


100. What is the scale factor for a 3/4" = 1'-0" ?

   A. 9
   B. 12
   C. 14
   D. 16

Answers to Review Questions A.

| | | | |
|---|---|---|---|
| 1. | C | 51. | G |
| 2. | D | 52. | E |
| 3. | C | 53. | C |
| 4. | A | 54. | B |
| 5. | D | 55. | A |
| 6. | D | 56. | C |
| 7. | B | 57. | B |
| 8. | D | 58. | D |
| 9. | A | 59. | B |
| 10. | B | 60. | B |
| 11. | B | 61. | A |
| 12. | D | 62. | B |
| 13. | C | 63. | A |
| 14. | B | 64. | A |
| 15. | D | 65. | B |
| 16. | C | 66. | A |
| 17. | B | 67. | C |
| 18. | C | 68. | D |
| 19. | A | 69. | B |
| 20. | B | 70. | C |
| 21. | C | 71. | B |
| 22. | B | 72. | C |
| 23. | C | 73. | A |
| 24. | C | 74. | A |
| 25. | A | 75. | D |
| 26. | F | 76. | C |
| 27. | A | 77. | D |
| 28. | C | 78. | D |
| 29. | B | 79. | B |
| 30. | H | 80. | D |
| 31. | E | 81. | B |
| 32. | G | 82. | D |
| 33. | D | 83. | C |
| 34. | D | 84. | C |
| 35. | E | 85. | B |
| 36. | A | 86. | B |
| 37. | C | 87. | D |
| 38. | B | 88. | D |
| 39. | D | 89. | D |
| 40. | C | 90. | B |
| 41. | A | 91. | C |
| 42. | E | 92. | B |
| 43. | B | 93. | A |
| 44. | H | 94. | D |
| 45. | D | 95. | C |
| 46. | B | 96. | D |
| 47. | I | 97. | D |
| 48. | A | 98. | C |
| 49. | F | 99. | B |
| 50. | J | 100. | D |

# REVIEW QUESTIONS B

Score:_____     Name:_____

MATCHING. Match ZOOM command options with their descriptions.

___1.   Calculates display magnification from the
        current display.

___2.   Causes entire drawing to be displayed on
        screen. This shows limits or if drawing
        size exceeds limits it also shows the
        entities of the drawing lying outside
        of the limits.

___3.   Shows all drawing entities at maxmium
        screen size.

___4.   Shows that portion of the drawing windowed
        at its maximum screen size.

___5.   Allows selection of a new display center point
        and a new magnification value; new display is
        centered on new center point.

___6.   Magnifies display based on lower left corner; X option is valid.

___7.   Provides an informational display showing current generated area,
        drawing extents, view box, and current view window.

___8.   Calculates amount of reduction or enlargement of display according
        to its intial display.

A.   ALL

B.   NUMBER (X)

C.   LEFT
     CORNER

D.   WINDOW

E.   EXTENTS

F.   NUMBER

G.   DYNAMIC

H.   CENTER

MATCHING. Match special characters with their TEXT code.

___9.   overscore                                    A.   %%d

___10.  underscore                                   B.   %%c

___11.  degrees symbol                               C.   %%p

___12.  plus/minus tolerance symbol                  D.   %%o

___13.  circle diameter dimensioning symbol          E.   %%u

MATCHING. Match distance measures available in UNITS with their format.

___14.  Scientific                                   A.   1'-9.75"

___15.  Decimal                                      B.   21 3/4"

___16.  Engineering                                  C.   21.75

___17.  Architectural                                D.   2.175E+1

___18.   Fractional                                  E.   1'-9 3/4"

145

MATCHING.     Match LAYER option with correct description.

___19. Provides listing of a defined layer including
        layer name, state, color, and linetype.

___20. Combines the actions of NEW and SET options.

___21. Makes the named layer current.

___22. Names the layer.

___23. Makes the named layer visible.

___24. Makes the named layer invisible.

___25. Defines a color for a chosen layer.

___26. Defines a linetype for the layer.

___27. Makes the layer invisible and prevents
        regeneration of the named layer.

___28. Makes the layer visible and enables regeneration of the named layer.

A.   ON

B.   SET

C.   THAW

D.   MAKE

E.   FREEZE

F.   OFF

G.   LTYPE

H.   ?

I.   NEW

J.   COLOR


MATCHING.   Match the View Ports Options.

___29. Lets you assign a name to the current configuration.

___30. Retrieves a previously saved viewport configuration.

___31. Erases a previously named viewport configuration.

___32. Divides the current view into two, three, or four
        viewports.

___33. Turns many viewports into one viewport.

___34. Displays the indentification number and screen
        positions of VPORTS.

___35. Merges two adjacent viewports into one larger
        viewport.

A.   ?

B.   2/3/4

C.   Single

D.   Restore

E.   Join

F.   Delete

G.   Save


MULTIPLE CHOICE.   Select the best answer.

36. The _____ command removes entities from the drawing.

        A. OOPS
        B. ERASE
        C. DELETE
        D. LAST

37. What command undoes the last erasure that was made?

        A. OOPS
        B. LAST
        C. UNERASE
        D. DARN!

38. Using the line command @ 3,4 is an example of a(n):

A. Absolute coordinate
B. Relative coordinate
C. Polar coordinate
D. Angular coordinate

39. The _____ command forces picked points to automatically align with an imaginary grid.

A. GRID
B. SNAP
C. OSNAP
D. AXIS

40. The _____ command places a ruler line on the edge of the screen.

A. GRID
B. SNAP
C. ORTHO
D. AXIS

41. The _____ command reverses several command moves at once.

A. U
B. UNDO
C. REDO
D. OOPS

42. The _____ command is used to erase part of a line, arc, circle, or trace.

A. ERASE
B. BREAK
C. TRIM
D. EXTEND

43. The _____ command is used to modify existing entities.

A. CHANGE
B. MOVE
C. COPY
D. MIRROR

44. The _____ command allows you to change the location of the entities and remain intact.

A. CHANGE
B. MOVE
C. COPY
D. MIRROR

45. A command that creates a 3-D wire frame plane.

A. 3DLINE
B. 3DMESH
C. 3DPLOY
D. 3DFACE

46. The _____ command is used to make multiple copies of one or more objects in rectangular or polar patterns.

    A. ARRAY
    B. MOVE
    C. COPY
    D. MINSERT

47. The _____ command allows you to change the scale of a(n) entity(ies). The X and Y scales are changed equally.

    A. ROTATE
    B. SCALE
    C. TRIM
    D. EXTEND

48. To bring up an attribute dialog box, which command would you use?

    A. ATTEXT
    B. ATTDEF
    C. ATTDISP
    D. ATTDIA

49. Which option of DVIEW allows you to change a 3-D drawing into a perspective?

    A. Target
    B. Zoom
    C. Camera
    D. Distance

50. The _____ command allows you to extend objects in a drawing to meet a boundary entity.

    A. ROTATE
    B. SCALE
    C. TRIM
    D. EXTEND

51. The _____ command causes AutoCAD to recalculate and redisplay the drawing database.

    A. REDRAW
    B. REGEN
    C. RECALC
    D. SCALE

52. The _____ command allows you to temporarily exit AutoCAD in order to use all DOS commands.

    A. EXIT
    B. SH
    C. SHELL
    D. END

53. The _____ command allows you to store a screen display identified by a name.

    A. BLOCK
    B. VIEW
    C. ZOOM P
    D. ID

54. A _____ is a pattern from which AutoCAD constructs text.

   A.  STYLE
   B.  QTEXT
   C.  FONT
   D.  DTEXT


55. How do you make line segments into a single ployline?

   A.  Change, Property, Pline, Join
   B.  Pline, Join
   C.  Pedit, Join
   D.  Change, Property, Thickness


56. What command is used to create a new text style?

   A.  TEXT, STYLE
   B.  TEXT
   C.  FONT
   D.  STYLE


57. If you enter 72%%% on the text line while in TEXT, what is generated on the screen?

   A.  72%%%
   B.  72<DIAMETER SIGN>
   C.  72<DEGREE SIGN>
   D.  72%


58. The _____ command allows the CAD user to set the drawing units appropriate for the type of drawing.

   A.  LIMITS
   B.  SCALE FACTOR
   C.  UNITS
   D.  TYPE


59. The basic definition of a layer **DOES NOT** include which of the following?

   A.  Layer name
   B.  Layer color
   C.  Layer linetype
   D.  Layer number


60. Which best describes the STATUS command?

   A.  Overrides the layer definition of a line type.
   B.  Causes new entities to be drawn in the color defined by the layer definition.
   C.  Rests the scale that defines how a line type is drawn.
   D.  Allows you to view the current drawing modes, defaults, limits, and other parameters in text form.


61. Which statement best describes the diameter option of the DIM command?

   A.  Draws an arc to show the angle between two lines.
   B.  Draws a dimension line across the diameter of the arc/circle.
   C.  Draws a dimension line like a radius in an arc or circle.
   D.  Draws only one dimension and then returns you to the command prompt.

62. Which of the following statements about LIMITS are TRUE?

A. Attempts to draw outside the preset boundaries are rejected.
B. Limits determine the area in which the grid is shown.
C. Limits determine the area ZOOM Extents will display.
D. A and B.

63. Whichof the following can not be changed when defining plot parameters?

A. Plot orgin
B. Hidden line removal
C. Pen numbers
D. Ltscale

64. Which LAYER option is matched with an incorrect description?

A. SET: Makes the named layer current.
B. NEW: Names the layer.
C. ON: Makes the named layer visible.
D. OFF: Makes the layer invisible and prevents regeneration of named layer.

65. Which statement best describes the LIST command?

A. Displays the absolute coordinates and current elevation of the chosen point.
B. Calculates the perimeter, in the set drawing units, for a closed polygon defined by user generated points.
C. Used to display the data stored by AutoCAD about any selected entity.
D. Causes all data stored by AutoCAD about each drawing entity to be listed.

66. Which statement best describes the DIVIDE command?

A. Displays absolute coordinates and current elevation of chosen point.
B. Calculates the perimeter, in the set drawing units, for a closed polygon defined by user generated points.
C. Separates an entity into an equal number of parts.
D. Divides an entity into intervals of specified lengths.

67. The BLOCK command:

A. Is a view of entities identified by a name.
B. Is used to identified a group of entities and to name them.
C. Takes entities defined as a block and writes them to the disk as a separate drawing.
D. Is used to define an insertion base point for a drawing file other than the default value: (0,0).

68. The HATCH command:

A. Is a pattern of lines, dots, and dashes that make up a defined pattern.
B. Is used to fill a specified polygon with a picked pattern using the chosen style
C. Allows freehand drawing in AutoCAD.
D. Is used to fill a polygon with a solid color.

69. To enter the Isometric Mode issue the _____ command.

A. GRID
B. ELLIPSE
C. ISOPLANE
D. SNAP

70. To change from one isometric axis to another, use

   A. CTRL E
   B. ISOPLANE command
   C. GRID
   D. A and B


71. Which of the following is FALSE?

   A. At the "Select Polyline" prompt in PEDIT, the line selected
      does not have to be a polyline.
   B. Vertex editing is used to modify vertices and segments
      between chosen vertices of a polyline.
   C. The DIVIDE command works only on lines, arcs, circles, or polylines.
   D. A block is not stored as part of the drawing file.


72. Which of the following is FALSE?

   A. The block insertion base point is the point chosen by the operator to
      define where the block is placed in reference to the block insertion point.
   B. Entities to be defined as a block can be chosen by pointing.
   C. When the BLOCK command is completed, the entities selected are erased
      from the screen.
   D. A block created on the O layer will, when inserted, be placed on the
      current layer.


73. Which of the following is FALSE?

   A. By issuing the BLOCK command and entering "?" to the prompt, all blocks
      associated with the drawing are listed.
   B. The INSERT command can be used to place a block, a file created by WBLOCK,
      or a drawing file in the current drawing.
   C. During block insertion, placing an asterisk (*) in front of the block name
      to be inserted has no effect.
   D. The EXPLODE command breaks a block into its component parts.


74. Which PLINE option and its description is incorrect?

   A. Close: Works like CLOSE command in LINE.
   B. Dist: Draws line segment at same angle as previous segment for specified length.
   C. Undo: Last polyline entered is undone.
   D. Width: Selects polyline width.


75. Which PEDIT option and its description is incorrect?

   A. Open: Converts and connects nonpolyline entities to polylines.
   B. Width: Allows selection of new width for all polyline segments.
   C. Fit Curve: Constructs a smooth curve using the vertices.
   D. X: Returns to command prompt.


76. The ERASE command

   A. Erases part of a entity.
   B. Erases any entity.
   C. Erases any group of entities.
   D. both B and C.

77. The OFFSET command

   A. Produces a mirror image of all objects.
   B. Contains a solid fill option.
   C. Allows creation of an entity parallel to another entity.
   D. None of the above.


78. AutoCAD text can be

   A. Left and right justified.
   B. Centered and aligned between two points.
   C. Created as large or as small as desired.
   D. All of the above.


79. Which one of the following commands is used to erase a small piece of a polyline?

   A. PEDIT
   B. ERASE
   C. UNDO
   D. BREAK


80. The common element of both the END and the QUIT command is that

   A. Both save your work.
   B. Neither saves your work.
   C. Both exit the Drawing Editor.
   D. Neither is useful.


81. The REVSURF command will

   A. Extrudes a curve segment in the direction and length you pick.
   B. Rotates a curve segment about an axis the number of degrees specified.
   C. Takes four segments and generates a mesh connecting them.
   D. Generates a surface between two entities.


82. Which DIM variable generates dimension limits?

   A. DIMTOL
   B. DIMTAD
   C. DIMTIM
   D. DIMLIM


83. Which DIM variable controls the space between the extension line from the object?

   A. DIMEXE
   B. DIMEXO
   C. DIMDLE
   D. DIMDLI


84. Which DIM variable changes the alternate units to read "cm" instead of "mm"?

   A. DIMALTF
   B. DIMLFAC
   C. DIMALT
   D. DIMRND

85. What is the proper order for making an attribute and then inserting it?

   A. Attdia, Insert, Block
   B. Attdisp, Block, Insert
   C. Attdef, Block, Insert
   D. Block, Attdia, Insert


86. Given 1/4" = 1'-0" what is the correct scale factor?

   A. .25
   B. 4
   C. 25
   D. 48


87. Given 1" = 20' what is the correct scale factor?

   A. .2
   B. 1
   C. 24
   D. 240


88. What command controls the degree of accuracy in displaying Arcs and Circles on the graphic screen?

   A. VIEW
   B. VIEWRES
   C. VIEWERS
   D. RESTORE


89. What command is used to set up and/or calibrate the Digitizer?

   A. Table
   B. Calibrate
   C. Configure
   D. Tablet


90. At the Command prompt which series of commands lists ALL Dimensioning Variables?

   A. Status
   B. List
   C. Dim, Status
   D. Dim, List


91. To obtain 1/8" dimension text height for a 3/4" = 1'-0", what should the DIMTXT be set to?

   A. 1/8"
   B. 2"
   C. 12"
   D. 16"


92. If your drawing scale is 1" = 10', what should the DIMSCALE be set to?

   A. 12
   B. 48
   C. 240
   D. 120

93. Which Tablet command is used to readjust the menu area?

   A. Setup
   B. Calibrate
   C. Configure
   D. Menu


94. What dim variable is used to extend the dimension line past the extension line?

   A. DIMDLI
   B. DIMDLE
   C. DIMEXO
   D. DIMEXE


95. To control the precision (number after the decimal point) of dimension text use.

   A. DIMRD
   B. ROUND
   C. UNITS
   D. ACCURACY


96. What is the scale factor for a 1" = 3'-0" ?

   A. .25
   B. 4
   C. 36
   D. 432


97. The ATTEDIT command is used on?

   A. DTEXT
   B. POLYGONS
   C. ATTRIBUTES
   D. PLINES


98. With the DIMCEN variable set to .09 what will you get when used.

   A. Center Mark and Center Lines
   B. Center Mark only
   C. Center Lines only
   D. A dot in the center of a circle


99. Which command WILL NOT work on a Polygon shape?

   A. EXPLODE
   B. PEDIT, "W" width option
   C. STRETCH, "C" option
   D. CHANGE, Endpoint option


100. Which command WILL work on Circles?

   A. EXPLODE
   B. PEDIT, "W" width option
   C. STRETCH, "C" option
   D. CHANGE, Endpoint option


154

Answers to Review Questions B.

| | | | |
|---|---|---|---|
| 1. | F | 51. | B |
| 2. | A | 52. | C |
| 3. | E | 53. | B |
| 4. | D | 54. | C |
| 5. | H | 55. | C |
| 6. | C | 56. | D |
| 7. | G | 57. | D |
| 8. | B | 58. | C |
| 9. | D | 59. | D |
| 10. | E | 60. | D |
| 11. | A | 61. | B |
| 12. | C | 62. | D |
| 13. | B | 63. | D |
| 14. | D | 64. | D |
| 15. | C | 65. | C |
| 16. | A | 66. | C |
| 17. | E | 67. | B |
| 18. | B | 68. | B |
| 19. | H | 69. | D |
| 20. | D | 70. | D |
| 21. | B | 71. | D |
| 22. | I | 72. | B |
| 23. | A | 73. | C |
| 24. | F | 74. | B |
| 25. | J | 75. | A |
| 26. | G | 76. | D |
| 27. | E | 77. | C |
| 28. | C | 78. | D |
| 29. | G | 79. | D |
| 30. | D | 80. | C |
| 31. | F | 81. | B |
| 32. | B | 82. | D |
| 33. | C | 83. | B |
| 34. | A | 84. | A |
| 35. | E | 85. | C |
| 36. | B | 86. | D |
| 37. | A | 87. | D |
| 38. | B | 88. | B |
| 39. | B | 89. | D |
| 40. | D | 90. | C |
| 41. | B | 91. | A |
| 42. | B | 92. | D |
| 43. | A | 93. | C |
| 44. | B | 94. | A |
| 45. | D | 95. | C |
| 46. | A | 96. | C |
| 47. | B | 97. | D |
| 48. | D | 98. | B |
| 49. | D | 99. | D |
| 50. | D | 100. | D |

PROBLEM A-7
GRAPHIC MATCHING

Using the choices provided, write the answer that describes the TEXT command option pictured. NOTE: "X" marks the pickpoint.

ALIGN, CENTER, FIT, LEFT MIDDLE, OR RIGHT

HOW
JUSTIFIED?

1. _____

HOW
JUSTIFIED?

2. _____

HOW
JUSTIFIED?

3. _____

HOW
JUSTIFIED?

4. _____

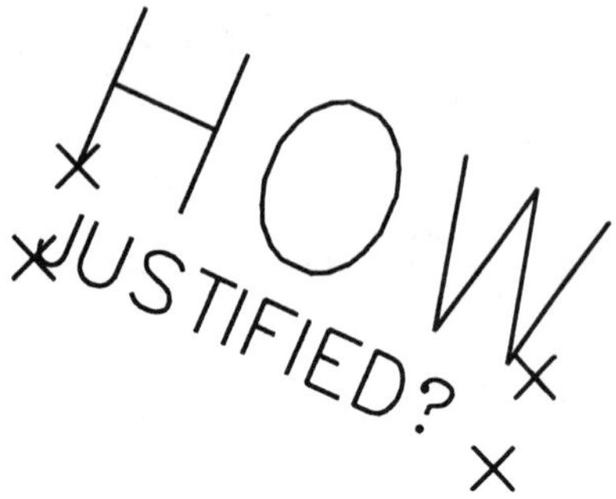

HOW
JUSTIFIED?

5. _____

HOW
JUSTIFIED?

6. _____

156

PROBLEM A-8

ARRAYING REVIEW

1. SHOW WHAT HAPPENS TO THE POLAR ARRAY WHEN IT IS ROTATED AND WHEN IT IS NOT, USE FOUR ITEMS, 360 DEGREES.

ROTATED                    NOT ROTATED

2. GIVEN THE ARRAY PATTERN, FILL IN THE DATA NECESSARY TO PRODUCE THE PATTERN.

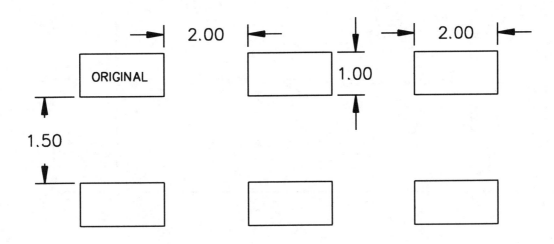

ARRAY
SELECT ITEMS: (BOX PICKED)
RECTANGULAR/POLAR: _____
NUMBER OF ROWS: _____
NUMBER OF COLUMNS: _____
CELL DIST. BETWEEN ROWS: _____
CELL DIST. BETWEEN COLUMNS: _____

# PROBLEM A-9
## MATCHING DIM VARIABLES

A. DIMTSZ
B. DIMASZ
C. DIMEXE
D. DIMEXO
E. DIMTAD
F. DIMSE1 (ON)
G. DIMSE2 (ON)
H. DIMSE1 (OFF)
I. DIMSE2 (OFF)
J. DIMTOD
K. DIMTXT
L. UNITS
M. DIMTOH
N. DIMTIH
O. DIMTIX
P. DIMDLE
Q. DIMDLI
R. DIMTXE

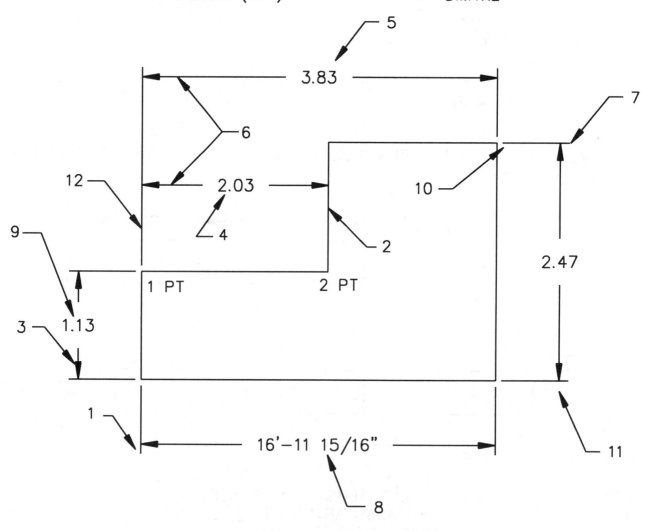

MATCH THE NUMBER TO THE CORRECT DIM VARIABLE.

1. _____
2. _____
3. _____
4. _____
5. _____
6. _____

7. _____
8. _____
9. _____
10. _____
11. _____
12. _____

# — NOTES —

# – NOTES –

# — NOTES —

# — NOTES —

# — NOTES —

# — NOTES —

# — NOTES —

# — NOTES —